AF564576

Ethics of Ecotourism

Ethics of Ecotourism

Dr. Parul Agarwal

RANDOM PUBLICATIONS
NEW DELHI (INDIA)

Ethics of Ecotourism

ISBN 978-93-5111-494-9

Published in 2015 in India by

RANDOM PUBLICATIONS

4376-A/4B, Gali Murari Lal, Ansari Road
New Delhi-110 002
Phone : +9111-43580356, 011-23289044, 011-43142548
e-mail: sales@randompublications.com,
info@randompublications.com, randomexports@gmail.com

Type Setting by : Friends Media, Delhi-110089
Digitally Printed at : Replika Press Pvt. Ltd.

Preface

Tourism is a global phenomenon with its characteristic as a large and complex business industry. Therefore, planning, developing and managing the tourism industry for a long-term success is a very difficult task. In order to achieve long-term success of tourism development, however, sustainable principles are required and necessary applied to strategic planning of tourism development processes. This involves various participations of all kinds of stakeholders from global to national and local levels respectively. To encourage the best practices of sustainable tourism development, voluntary initiative and codes of conduct are considerably as an effective tool on preventing or reducing negative impacts caused by tourists.

Based on this, the present book focused mainly on the implementation of voluntary initiatives and codes of conduct in the process of achieving sustainable tourism development in ecotourism context. The introduction of these measures aims to protect tourism resources in terms of natural, cultural, historical and other resources from the increasing numbers of tourists in a tourist destination. Consequently, the negative impacts of tourists on the destination can be minimized by the way of educating and changing their behavior with sustainable approach. As the result, tourism industry can be kept as a main driver for economic growth of a country, the destination competitiveness is enhanced and other positive benefits are maximized while the negative impacts are minimized.

This book examines the burgeoning industry of ecotourism, analyzing denitions of "ecotourism" and exploring a number of compelling issues raised by the recent trend in worldwide tourism. It also examines the international codes of ecotourism, their presuppositions, values and limitations. Far from exhausting the possibilities, this book opens up the connections between the ethical approaches and ecotourism.

Author

Contents

	Preface	**v**
1.	**Characteristics of Ecotourism**	**1**
	The Concept of Ecotourism	3
	Historical Background	5
	Environmental Impacts of Ecotourism	7
	Socio-economic Impacts of Ecotourism	10
	Regulation of Ecotourism	11
	Ecotourism and Ethics	14
	Codes of Conduct for Ecotourism	15
	Frameworks for Negotiating Discourses about Ecotourism	19
2.	**Global Code of Ethics for Tourism**	**26**
	Article 1. Tourism's Contribution to Mutual Understanding and Respect between Peoples and Societies	27
	Article 2. Tourism as a Vehicle for Individual and Collective Fulfilment	28
	Article 3. Tourism, a Factor of Sustainable Development	28
	Article 4. Tourism, a User of the Cultural Heritage of Mankind and Contributor to its Enhancement	29
	Article 5. Tourism, a Beneficial Activity for Host Countries and Communities	30
	Article 6. Obligations of Stakeholders in Tourism Development	30
	Article 7. Right to Tourism	32
	Article 8. Liberty of Tourist Movements	32
	Article 9. Rights of the Workers and Entrepreneurs in the Tourism Industry	33
	Article 10. Implementation of the Principles of the Global Code of Ethics for Tourism	34

3. Québec Declaration on Ecotourism **35**

A. To National, Regional and Local Governments 38

B. To the Private Sector 41

C. To Non-governmental Organisations, Community-based Associations, Academic and Research Institutions 43

D. To Inter-governmental Organisations, International Financial Institutions and Development Assistance Agencies 43

E. To Local and Indigenous Communities 45

F. To the World Summit on Sustainable Development (WSSD) 45

4. Ecotourism: Planning and Management **46**

Ecotourism Participants 47

Role of Private and Public Sectors 53

Finding a Proper Balance 56

Governments' Role in Decision-Making 58

Ecotourism Management Plan (EMP) 59

5. Guidelines for Sustainable Ecotourism **104**

Working with Ecotourism 107

Sustainability in the Tourism Sector 109

Factors Affecting the Development of Ecotourism 111

UNEP Principles of Sustainable Tourism 120

Charter of Sustainable Tourism 128

Berlin Declaration on Biodiversity and Tourism 132

Responsible Tourism in Destinations: The Cape Town Declaration, 2002 138

6. Ecotourism and Biodiversity Conservation **145**

Values of Biodiversity 147

Ecotourism and Biodiversity Conservation Planning 153

Sustainable Tourism Industry 163

Conventional Mass Tourism 165

Environmental Carrying Capacity 169

Tourism as a Multi-sectoral Phenomenon 170

Certification System 174

Biodiversity and Nature-based Tourism 176

International Convention on Biological Diversity 180
Protection of Natural Areas and Biodiversity 181
Strategic Planning For NBE Development 190

7. Ecotourism in Protected Areas 195

The Role of Ecotourism 196
Opportunities and Threats 198
Potential Opportunities of Ecotourism 199
Issues of Tourism in Protected Areas 205
Survival of Protected Areas 209
Tourism Markets and Protected Areas 212
Managing Tourism in Protected Areas 214
Structures for Protected Area Management 218
Funding Methods for Protected Areas 224
Trends in Park Tourism 231
Park Tourism: Planning and Management Issues 237

8. International Ecotourism Initiatives 239

Ecotourism and Question of "Sustainability" 239
Ecotourism as Development 241
Globalisation and Liberalisation 242
Environmental trap 243
Social Erosion 247
Indigenous Cultures Threatened 249
Legitimising Discourse of Ecotourism 251
Policy Proposals 253
International Year of Ecotourism 255
Initiative by World Tourism Organisation 256
Ecotourism Policies 275

Bibliography 285

Index 287

1

Characteristics of Ecotourism

According to the Quebec Declaration on Ecotourism, ecotourism "embraces the principles of sustainable tourism... and the following principles which distinguish it from the wider concept of sustainable tourism:

- Contributes actively to the conservation of natural and cultural heritage,
- Includes local and indigenous communities in its planning, development and operation, contributing to their well-being,
- Interprets the natural and cultural heritage of the destination to visitor,
- Lends itself better to independent travelers, as well as to organized tours for small size groups".

As a development tool, ecotourism can advance the three basic goals of the Convention on Biological Diversity which are:

- The conservation of biological and cultural diversity, by strengthening protected area management and increasing the value of ecosystems;
- The promotion of the sustainable use of biodiversity, by generating income, jobs and business opportunities in ecotourism and related business networks, and
- Sharing the benefits of ecotourism developments equitably with local communities and indigenous people, by obtaining their informed consent and full participation in planning and management of ecotourism businesses.

In the field, well-planned and managed ecotourism has proven to be one of the most effective tools for long-term conservation of biodiversity when the right circumstances (such as market feasibility, management capacity at local level, and clear and monitored links between ecotourism development and conservation) are present. In May 2000, as part of the side events on the 8th session of the United Nations Commission on Sustainable Development (CSD 8), a group of Indigenous Peoples Organizations, NGOs and other members of Civil Society provided a proposal on guidelines for ecotourism. Although the final result could not be incorporated into the official papers due to procedural aspects, UNEP recognizes its value as a statement of genuine concerns from primary stakeholders.

Ecotourism is sustainable tourism, which follows clear processes that:

- Ensures prior informed participation of all stakeholders,
- Ensures equal, effective and active participation of all stakeholders,
- Acknowledges Indigenous Peoples communities' rights to say "no" to tourism development - and to be fully informed, effective and active participants in the development of tourism activities within the communities, lands, and territories, and
- Promotes processes for Indigenous Peoples and local communities to control and maintain their resources.

In July 1998 the United Nations Economic and Social Council (ECOSOC) proposed to members of the UN General Assembly to designate 2002 as the International Year of Ecotourism (IYE).

The members of ECOSOC consider the designation of the IYE as an encouragement for intensified cooperative efforts by Governments and international and regional organizations, as well as non-governmental organizations, to achieve the aims of Agenda 21 in promoting development and the protection of the environment.

Recognizing the growing importance of ecotourism, the UN General Assembly in December 1998 accepted the proposal and declared 2002 as the International Year of Ecotourism. The Commission on Sustainable Development (CSD) and other venues were advised to implement the Year. Within the UN system the CSD's Interagency Committee on Sustainable Development (IACSD) mandated the World Tourism Organization (WTO/OMT) and UNEP to prepare and coordinate supportive activities for and during the year.

Activities supporting the IYE 2002 and the World Ecotourism Summit in Quebec in May 2002 were conducted as part of UNEP's Tourism Programme mission to ensure that the sustainable use and management of the natural, cultural and man-made environment is an integral part of all tourism development. The 2002 events were carried out through partnerships with the World Tourism Organization (WTO/OMT), The International Ecotourism Society (TIES), Ecological Tourism in Europe (ETE), and others.

The Concept of Ecotourism

Ecotourism has been developed following the environmental movement which appeared at the beginning of the seventies. The growing interest of people for environment and trips oriented towards fresh air, in addition to the growing dissatisfaction towards mass tourism, highlighted to the tourism industry a need for ecotourism. Besides, the understanding and the agreement with the principles of nature preservation and durability for a growing portion of the population took part in the evolution of the term "ecotourism".

Budowski is usually cited as the pioneer for the ecotourism concept. In his article Tourism and Environmental Conservation: Conflict, Coexistance or Symbiosis?, Budowski admits that the relation between tourism and natural environment tends to be in conflict, but believes that there exists a possibility for a relation based upon mutual benefits. His description of the possible symbiotic relation that could follow from this corresponds to our contemporary idea of ecotourism.

Definitions for Ecotourism

Ecotourism is often considered as a form of tourism with "a strong motivation". There is no universal definition for ecotourism. It is usually considered as a "tourism favourable to the environment", which is, on a practical level, variously interpreted according to the country.

In the absence of a clear and recognized definition, the definition for the International Society for Ecotourism (1991) is: "... a responsible tourism in natural environment which preserves it and participates to the well-being of local populations".

According to the World Conservation Union (1996), it can be defined as "... the visit of natural environments remained relatively intact... with a low negative impact... including a socio-economical implication for the local populations which is at the same time active and beneficial".

Although it is difficult to define ecotourism, it presents several characteristics:

- the destination is generally a natural environment which is not polluted;
- its attractions are its flora and its wildlife, and more generally it bio-diversity;
- ecotourism must support the local economy and the specificity of the place;
- it must contribute to the preservation of the environment, and more generally, promote the preservation of nature;
- ecotourist stays often include an educational aspect.

Lasting tourism or ecotourism?

According to the definition and principles of ecotourism established by The International Ecotourism Society (TIES) in 1990, ecotourism is "Responsible travel to natural areas that conserves the environment and improves the well-being of local people.". Martha Honey, expands on the TIES definition by describing the seven characteristics of ecotourism, which are:

1. Involves travel to natural destinations
2. Minimises impact.
3. Builds environmental awareness.
4. Provides direct financial benefits for conservation.
5. Provides financial benefits and empowerment for local people.
6. Respects local culture.
7. Supports human rights and democratic movements such as:
   - conservation of biological diversity and cultural diversity through ecosystem protection.
   - promotion of sustainable use of biodiversity, by providing jobs to local populations.
   - sharing of socio-economic benefits with local communities and indigenous peoples by having their informed consent and participation in the management of ecotourism enterprises.
   - tourism to unspoiled natural resources, with minimal impact on the environment being a primary concern.

— minimisation of tourism's own environmental impact.
— affordability and lack of waste in the form of luxury.
— local culture, flora and fauna being the main attractions.
— local people benefit from this form of tourism economically, often more than mass tourism.

Ecotourism should not be confounded with lasting tourism. The former is a type of tourism (such as sport, cultural, leisure, or adventure tourism), whereas the latter refers to the way tourism must be applied to all of these types of tourism. If the principles of lasting tourism are applied, all of theses types of tourism may be defined as lasting.

Whereas the first definitions of ecotourism stressed the proximity with nature that tourists looked for, more recent definitions aimed at placing the accent on a variety of principles associated to the concept of lasting tourism. Nowadays, we commonly admit that ecotourism includes the principles of lasting tourism regarding the impact of such an activity on economy, society, and environment, and additionally, it includes the following specific principles which distinguish it from the broader concept of lasting tourism:

- ecotourism actively contributes to the natural and cultural heritage protection;
- ecotourism includes local and native populations in its planning, development, and exploitation, and it contributes to their well-being;
- ecotourism offers to visitors an interpretation of the natural and cultural heritage;
- ecotourism lends itself better to individual travelling and travelling organized for small groups.

To summarize, an analysis of definitions brings us to consider three dimensions which characterize the essence of ecotourism:

- a tourism axed on nature;
- an educational aspect;
- a need for durability.

Historical Background

Hector Ceballos-Lascurain popularised the term 'ecotourism' in July 1983, when he was performing the dual role of Director General of Standards and Technology of SEDUE (the Mexican Ministry of Urban Development and

Ecology) and founding president of PRONATURA (an influential Mexican conservationist NGO). PRONATURA was lobbying for the conservation of the wetlands in northern Yucatán as breeding and feeding habitats of the American Flamingo. Others claim the term was in use earlier: Claus-Dieter (Nick) Hetzer, an academic and adventurer from Forum International in Berkeley, CA, coined the term in 1965 and ran the first ecotours in the Yucatán during the early 1970s.

Ecotourism and sustainable development have become prevalent concepts since the mid 1980s, and ecotourism has experienced arguably the fastest growth of all sub-sectors in the tourism industry. The popularity represents a change in tourist perceptions, increased environmental awareness, and a desire to explore natural environments. At times, such changes become as much a statement affirming one's social identity, educational sophistication, and disposable income as it has about preserving the Amazon rainforest or the Caribbean reef for posterity. However, in the continuum of tourism activities that stretch from conventional tourism to ecotourism proper, there has been a lot of contention to the limit at which biodiversity preservation, local social-economic benefits, and environmental impact can be considered "ecotourism". For this reason, environmentalists, special interest groups, and governments define ecotourism differently.

Environmental organisations have generally insisted that ecotourism is nature-based, sustainably managed, conservation supporting, and environmentally educated. The tourist industry and governments, however, focus more on the product aspect, treating ecotourism as equivalent to any sort of tourism based in nature. As a further complication, many terms are used under the rubric of ecotourism. Nature tourism, low impact tourism, green tourism, bio-tourism, ecologically responsible tourism, and others have been used in literature and marketing, although they are not necessary synonymous with ecotourism.

The problems associated with defining ecotourism have led to confusion among tourists and academics. Definitional problems are also subject of considerable public controversy and concern because of green washing, a trend towards the commercialisation of tourism schemes disguised as sustainable, nature based, and environmentally friendly ecotourism. According to McLaren, these schemes are environmentally destructive, economically exploitative, and culturally insensitive at its worst. They are also morally disconcerting because they mislead tourists and manipulate their concerns for the environment. The development and success of such large scale, energy intensive, and ecologically unsustainable schemes are a testament to the tremendous profits associated with being labelled as ecotourism.

Environmental Impacts of Ecotourism

Ecotourism has become one of the fastest-growing sectors of the tourism industry, growing annually by 10-15% worldwide. One definition of ecotourism is "the practice of low-impact, educational, ecologically and culturally sensitive travel that benefits local communities and host countries". Many of the ecotourism projects are not meeting these standards. Even if some of the guidelines are being executed, the local communities are still facing other negative impacts. South Africa is one of the countries that are reaping significant economic benefits from ecotourism, but negative effects - including forcing people to leave their homes, gross violations of fundamental rights, and environmental hazards - far outweigh the medium-term economic benefits.

A tremendous amount of money is being spent and human resources continue to be used for ecotourism despite unsuccessful outcomes, and even more money is put into public relation campaigns to dilute the effects of criticism. Ecotourism channels resources away from other projects that could contribute more sustainable and realistic solutions to pressing social and environmental problems. "The money tourism can generate often ties parks and managements to ecotourism". But there is a tension in this relationship because ecotourism often causes conflict and changes in land-use rights, fails to deliver promises of community-level benefits, damages environments, and has plenty of other social impacts. Indeed many argue repeatedly that ecotourism is neither ecologically nor socially beneficial, yet it persists as a strategy for conservation and development. While several studies are being done on ways to improve the ecotourism structure, some argue that these examples provide rationale for stopping it altogether.

The ecotourism system exercises tremendous financial and political influence. The evidence above shows that a strong case exists for restraining such activities in certain locations. Funding could be used for field studies aimed at finding alternative solutions to tourism and the diverse problems Africa faces in result of urbanisation, industrialisation, and the over exploitation of agriculture. At the local level, ecotourism has become a source of conflict over control of land, resources, and tourism profits. In this case, ecotourism has harmed the environment and local people, and has led to conflicts over profit distribution. In a perfect world more efforts would be made towards educating tourists of the environmental and social effects of their travels. Very few regulations or laws stand in place as boundaries for the investors in ecotourism. These should be implemented to prohibit the promotion of unsustainable ecotourism projects and materials which project false images of destinations, demeaning local and indigenous cultures.

Ecotourism operations occasionally fail to live up to conservation ideals. It is sometimes overlooked that ecotourism is a highly consumer-centred activity, and that environmental conservation is a means to further economic growth. Although ecotourism is intended for small groups, even a modest increase in population, however temporary, puts extra pressure on the local environment and necessitates the development of additional infrastructure and amenities. The construction of water treatment plants, sanitation facilities, and lodges come with the exploitation of non-renewable energy sources and the utilisation of already limited local resources. The conversion of natural land to such tourist infrastructure is implicated in deforestation and habitat deterioration of butterflies in Mexico and squirrel monkeys in Costa Rica. In other cases, the environment suffers because local communities are unable to meet the infrastructure demands of ecotourism. The lack of adequate sanitation facilities in many East African parks results in the disposal of campsite sewage in rivers, contaminating the wildlife, livestock, and people who draw drinking water from it.

Aside from environmental degradation with tourist infrastructure, population pressures from ecotourism also leaves behind garbage and pollution associated with the Western lifestyle. Although ecotourists claim to be educationally sophisticated and environmentally concerned, they rarely understand the ecological consequences of their visits and how their day-to-day activities append physical impacts on the environment. As one scientist observes, they "rarely acknowledge how the meals they eat, the toilets they flush, the water they drink, and so on, are all part of broader regional economic and ecological systems they are helping to reconfigure with their very activities." Nor do ecotourists recognise the great consumption of non-renewable energy required to arrive at their destination, which is typically more remote than conventional tourism destinations. For instance, an exotic journey to a place 10,000 kilometers away consumes about 700 liters of fuel per person.

Ecotourism activities are, in of itself, issues in environmental impact because they disturb fauna and flora. Ecotourists believe that because they are only taking pictures and leaving footprints, they keep ecotourism sites pristine, but even harmless sounding activities such as a nature hike can be ecologically destructive. In the Annapurna Circuit in Nepal, ecotourists have worn down the marked trails and created alternate routes, contributing to soil impaction, erosion, and plant damage. Where the ecotourism activity involves wildlife viewing, it can scare away animals, disrupt their feeding and nesting sites, or acclimate them to the presence of people. In Kenya, wildlife-observer disruption drives cheetahs off their reserves, increasing the risk of inbreeding and further endangering the species.

The industrialisation, urbanisation, and unsustainable agriculture practices of human society are considered to be having a serious effect on the environment. Ecotourism is now also considered to be playing a role in this depletion. While the term ecotourism may sound relatively benign, one of its most serious impacts is its consumption of virgin territories. These invasions often include deforestation, disruption of ecological life systems and various forms of pollution, all of which contribute to environmental degradation. The number of motor vehicles crossing the park increases as tour drivers search for rare species.

The number of roads has disrupted the grass cover which has serious effects on plant and animal species. These areas also have a higher rate of disturbances and invasive species because of all the traffic moving off the beaten path into new undiscovered areas. Ecotourism also has an effect on species through the value placed on them. "Certain species have gone from being little known or valued by local people to being highly valued commodities. The commodification of plants may erase their social value and lead to overproduction within protected areas. Local people and their images can also be turned into commodities".

Kamuaro brings up a relatively obvious contradiction, any commercial venture into unspoiled, pristine land with or without the "eco" prefix as a contradiction in terms. To generate revenue you have to have a high number of traffic, tourists, which inevitably means a higher pressure on the environment.

Socio-economic Impacts of Ecotourism

Most forms of ecotourism are owned by foreign investors and corporations that provide few benefits to local communities. An overwhelming majority of profits are put into the pockets of investors instead of reinvestment into the local economy or environmental protection. The limited numbers of local people who are employed in the economy enter at its lowest level, and are unable to live in tourist areas because of meager wages and a two market system.

In some cases, the resentment by local people results in environmental degradation. As a highly publicised case, the Masai nomads in Kenya killed wildlife in national parks to show aversion to unfair compensation terms and displacement from traditional lands. The lack of economic opportunities for local people also constrains them to degrade the environment as a means of sustenance. The presence of affluent ecotourists encourage the development of destructive markets in wildlife souvenirs, such as the sale of coral trinkets on tropical islands and animal products in Asia, contributing to illegal harvesting and poaching from the environment. In Suriname, sea turtle reserves use a very large portion of their budget to guard against these destructive activities.

One of the most powerful examples of communities being moved in order to create a park is the story of the Masai. About 70% of national parks and game reserves in East Africa are on Masai land. The first undesirable impact of tourism was that of the extent of land lost from the Masai culture. Local and national governments took advantage of the Masai's ignorance on the situation and robbed them of huge chunks of grazing land, putting to risk their only socio-economic livelihood. In Kenya the Masai also have not gained any economic benefits. Despite the loss of their land, employment favours better educated workers. Furthermore the investors in this area are not local and have not put profits back into local economy. In some cases game reserves can be created without informing or consulting local people, who come to find out about the situation when an eviction notice is delivered. Another source of resentment is the manipulation of the local people by their government.

Ecotourism works to create simplistic images of local people and their uses and understandings of their surroundings. Through the lens of these simplified images, officials direct policies and projects towards the local people and the local people are blamed if the projects fail. Clearly tourism as a trade is not empowering the local people who make it rich and satisfying. Instead ecotourism exploits and depletes, particularly in African Masai tribes. It has to be reoriented if it is to be useful to local communities and to become sustainable.

Ecotourism often claims that it preserves and "enhances" local cultures. However, evidence shows that with the establishment of protected areas local people have illegally lost their homes, and most often with no compensation. Pushing people onto marginal lands with harsh climates, poor soils, lack of water, and infested with livestock and disease does little to enhance livelihoods even when a proportion of ecotourism profits are directed back into the community. The establishment of parks can create harsh survival realities and deprive the people of their traditional use of land and natural resources. Ethnic groups are increasingly being seen as a "backdrop" to the scenery and wildlife. The local people struggle for cultural survival and freedom of cultural expression while being "observed" by tourists. Local indigenous people also have strong resentment towards the change, "Tourism has been allowed to develop with virtually no controls. Too many lodges have been built, too much firewood is being used and no limits are being placed on tourism vehicles. They regularly drive off-track and harass the wildlife. Their vehicle tracks criss-cross the entire Masai Mara. Inevitably the bush is becoming eroded and degraded".

Regulation of Ecotourism

Because the regulation of ecotourism is poorly implemented or nonexistent, ecologically destructive green washed operations like underwater hotels, helicopter tours, and wildlife theme parks are categorised as ecotourism along with canoeing, camping, photography, and wildlife observation. The failure to acknowledge responsible, low impact ecotourism puts these companies at a competitive disadvantage.

Many environmentalists have argued for a global standard of accreditation, differentiating ecotourism companies based on their level of environmental commitment. A national or international regulatory board would enforce accreditation procedures, with representation from various groups including governments, hotels, tour operators, travel agents, guides, airlines, local authorities, conservation organisations, and non-governmental organisations. The decisions of the board would be sanctioned by governments, so that non-compliant companies would be legally required to disassociate themselves from the use of the ecotourism brand.

Crinion suggests a Green Stars System, based on criteria including a management plan, benefit for the local community, small group interaction, education value and staff training. Ecotourists who consider their choices would be confident of a genuine ecotourism experience when they see the higher star rating.

In addition, environmental impact assessments could be used as a form of accreditation. Feasibility is evaluated from a scientific basis, and recommendations could be made to optimally plan infrastructure, set tourist capacity, and manage the ecology. This form of accreditation is more sensitive to site specific conditions.

An environmental protection strategy must address the issue of ecotourists removed from the cause-and-effect of their actions on the environment. More initiatives should be carried out to improve their awareness, sensitise them to environmental issues, and care about the places they visit. Tour guides are an obvious and direct medium to communicate awareness. With the confidence of ecotourists and intimate knowledge of the environment, they can actively discuss conservation issues. A tour guide training program in Costa Rica's Tortuguero National Park has helped mitigate negative environmental impacts by providing information and regulating tourists on the parks' beaches used by nesting endangered sea turtles.

The underdevelopment theory of tourism describes a new form of imperialism by multinational corporations that control ecotourism resources. These corporations finance and profit from the development of large scale ecotourism that causes excessive environmental degradation, loss of traditional culture and way of life, and exploitation of local labor. In Zimbabwe and Nepal's Annapurna region, where underdevelopment is taking place, more than 90 percent of ecotourism revenues are expatriated to the parent countries, and less than 5 percent go into local communities.

The lack of sustainability highlights the need for small scale, slow growth, and locally based ecotourism. Local peoples have a vested interest in the well being of their community, and are therefore more accountable to environmental protection than multinational corporations. The lack of control, westernisation, adverse impacts to the environment, loss of culture and traditions outweigh the benefits of establishing large scale ecotourism.

The increased contributions of communities to locally managed ecotourism create viable economic opportunities, including high level management positions, and reduce environmental issues associated with poverty and unemployment. Because the ecotourism experience is marketed to a different lifestyle from large scale ecotourism, the development of facilities and infrastructure does not need to conform to corporate Western tourism standards, and can be much simpler and less expensive. There is a greater multiplier effect on the economy, because local products, materials, and labor are used. Profits accrue locally and import leakages are reduced. However, even this form of tourism may require foreign investment for promotion or start up. When such investments are required, it is crucial for communities for find a company or non-governmental organisation that reflects the philosophy of ecotourism; sensitive to their concerns and willing to cooperate at the expense of profit. The basic assumption of the multiplier effect is that the economy starts off with unused resources, for example, that many workers are cyclically unemployed and much of industrial capacity is sitting idle or incompletely utilised. By increasing demand in the economy it is then possible to boost production. If the economy was already at full employment, with only structural, frictional, or other supply-side types of unemployment, any attempt to boost demand would only lead to inflation. For various laissez-faire schools of economics which embrace Say's Law and deny the possibility of Keynesian inefficiency and under-employment of resources, therefore, the multiplier concept is irrelevant or wrong-headed.

As an example, consider the government increasing its expenditure on roads by $1 million, without a corresponding increase in taxation. This sum would go to the road builders, who would hire more workers and distribute the money as wages and profits. The households receiving these incomes will save part of the money and spend the rest on consumer goods. These expenditures in turn will generate more jobs, wages, and profits, and so on with the income and spending circulating around the economy.

The multiplier effect arises because of the induced increases in consumer spending which occur due to the increased incomes — and because of the feedback into increasing business revenues, jobs, and income again. This process does not lead to an economic explosion not only because of the supply-side barriers at potential output (full employment) but because at each "round", the increase in consumer spending is less than the increase in consumer incomes. That is, the marginal propensity to consume (mpc) is less than one, so that each round some extra income goes into saving, leaking out of the cumulative process. Each increase in spending is thus smaller than that of the previous round, preventing an explosion. Ecotourism has to be implemented with care.

Natural resource management can be utilised as a specialised tool for the development of ecotourism. There are several places throughout the world where the amount of natural resources are abundant. But, with human encroachment and habitats these resources are depleting. Without knowing the proper utilisation of certain resources they are destroyed and floral and faunal species are becoming extinct. Ecotourism programmes can be introduced for the conservation of these resources. Several plans and proper management programmes can be introduced so that these resources remain untouched. Several organisations, NGO's, scientists are working on this field.

Ecotourism and Ethics

The use of the term itself-ecotourism-is not without problems. The literature devoted to ecotourism reflects these problems. In fact, the problems run the gamut from those ecotours that genuinely respect the environment and local populations of the host area to the marketing ploy that simply re-packages the old holiday tour with the term "eco" tacked on in front in order to exploit public sensibility about the environment. There is no doubt that worldwide public opinion is becoming increasingly concerned with environmental problems. The nature of these concerns differs enormously around the globe depending on complex economic, social, and cultural conditions. Global issues of access, use, and exploitation of limited resources as well as uneven and inequitable distribution of goods and services greatly complicate ecotourism and adventure travel. In the midst of this, it is important to raise the question of how an environmentally concerned traveler can make informed judgments about the marketing claims of travel packages that use the label "ecotourism" and purport to be environmentally sound, economically helpful to the local community, and ethical to boot.

In her article "Environmentally Responsible Marketing of Tourism," Pamela Wight goes on to describe the two prevailing views of ecotourism:

> One envisages that public interest in the environment may be used to market a product; the other sees that this same interest may be used to conserve the resources upon which the product is based. These views need not be mutually exclusive and may very well be complementary.

Sustainable tourism as a subset of sustainable development: since the 1987 publication of the report from the UN World Commission on the Environment and Development (also known as the Brundtland Commission), the concept of sustainable development has been moving more fully into political and economic discourse. By and large, this is a good thing, but it also raises important questions. The brief definition of sustainable development is that it is the process by which the needs of the present are met without compromising the ability of future generations to meet their own needs (WCED 8). Michael Redclift highlights the fact that the Brundtland Commission report emphasizes human needs, but does not give due weight to protecting the environment. Even more critical are William Rees's observations that the very way of thinking that views nature solely in material and mechanistic terms cannot get us out of the deep problems we have created in the first place because of these very attitudes toward nature. Nothing less than a paradigm shift in our thinking is called for in order to effect meaningful change. Nevertheless, even given these points, the notion of sustainable tourism can be a powerful heuristic for thinking through environmental issues related to tourism on the local level. Moreover, it would make sense to work out some of the problems of sustainable tourism on the local level using ecotourism and adventure travel as experiments to see if sustainable tourism can be made to work.

Codes of Conduct for Ecotourism

Even the most benighted tour operators realize that it makes no sense to foul the nest while at the same time they hope to feather it. Public opinion is highly sensitive to changing conditions at tourist destinations and it does not take long in this Internet era for word to get out that the reality on the ground fails to live up to the marketing rhetoric in the glossy brochures. This is especially the case with those who have the interest, resources, and leisure time to go on ecotours and adventure travels. Increasingly over the last decade or so, stakeholders in these types of tourism have been meeting and working to ensure a quality experience for travelers with a minimum of harmful effects on the local ecosystems and human communities. This balancing act has not been easy to achieve, especially with ever-increasing numbers of people who want to travel to "exotic" places around the world. Furthermore, the drive for financial gain cannot be underestimated both for the tour operators as well as for the host communities. These problems are even more pronounced in the developing world in which more traditional ways of life as well as local resource bases are made more vulnerable by possible conflicts between short-term use by local populations and the attractions of tourist monies flowing into local economies. Clearly, systemic and structural responses are needed to address these issues if long-term, sustainable and equitable solutions are to be found. Moreover, these issues need responses that are multi-faceted and multi-disciplinary in methodology and include economics, environmental studies, public policy, marketing, grass-roots collaboration, anthropology, sociology and finally ethics and applied ethics. In much of the literature, at least some nod is given to ethics as it relates to these pressing issues, but it often appears in the conclusion of the articles where authors mention ethics under the rubric of what needs further examination or study.

Much of the discussion about tourism in general and even ecotourism uses arguments based on the view of nature as a resource or a commodity. This view is highly anthropocentric and assumes as its starting points the notions of human dominion over nature and/or the fundamental divide between human beings and the rest of nature. These assumptions themselves need to be unpacked and analyzed in any full treatment of the issues facing ecotourism and adventure travel. Furthermore, many of the arguments are also based on some version of enlightened self-interest whether on the part of the tour operators, host communities or tourists themselves.

Essentially, these arguments take the form of an injunction to take care of the natural site so that it will continue to be an attractive tour destination. The chairman of the Association of Independent Tour Operators in Europe put it this way, "… it is time to protect the product on which our businesses depend … if we do not help conserve the very places which our clients clamour to see then, in five years, we shall have no clients at all… those dream locations will have been ruined." This is not to say that either ecotour managers or ecotourists themselves operate exclusively from anthropocentric norms or that they do not operate from the recognition of the inherent value of the places they visit. It is rather to claim that the arguments made for the preservation of specific locations (especially in the ecotourism literature and advertising) often appeal to self-regarding and anthropocentric norms, that is, that these locations be preserved so that human beings will have the opportunity to experience and appreciate them.

In recognition of and as a response to some problems raised by ecotourism, codes of ethics and codes of conduct have been developed by and for tour operators, nature organizations, ecotourists, and host communities. Here, three of these codes are discussed. It is revelatory to examine these codes to see what is being valued in them, how the maxims are stated and what their limitation are. The fact that they are three different kinds of codes illustrates that there are various ways to approach ecotourism: one code is site-specific (Antarctica); another comes from a leading nature group (the National Audubon Society); and the third has been developed by one of the leading figures in travel research, Stanley Plog who heads a consulting firm for travel research and is an editor for the Journal of Travel Research. We look at four maxims that are somewhat similar and that are gleaned from the three codes: one each from Audubon and Plog and two from the Antarctic Code.

Antarctic Code

1. Antarctic visitors must not leave footprints in fragile mosses, lichens or grasses.
2. Antarctic visitors must not violate the seals', penguins' or seabirds' personal space.
   - Start with a "baseline" distance of 15 ft. (5 m) from penguins, seabirds, true seals, and 60 ft. (18 m) from fur seals.
   - Give the animals the right-of-way.
   - Stay on the edge of, and don't walk through, animal groups.
   - Back off if necessary.
   - Never touch the animals.

Audubon travel ethic:

3. Audubon tours must strengthen the conservation effort and enhance the integrity of the places visited.

The major concern of these three maxims is the integrity and enhancement of a specific site as well as the protection of particular animals at the location. Moreover, the Audubon maxim foregrounds the potential effects of the tour itself and seeks to ally the ecotour with those forces that "strengthen the conservation effort." This last point goes well beyond doing no harm or having zero-impact on a site and commits the tour to assisting the conservation effort. Of course, no particulars are given to show how this might occur with specific tours, but it is clear that the Audubon Society wants to do more than the bare minimum that might be expected from any tour, namely that the host communities not be harmed. It would be interesting to follow up on some recent Audubon tours to see if and how efforts have been made to achieve this goal.

The Audubon Society's injunction to augment conservation efforts and enhance "natural integrity" apart from any instrumental value to human beings is in harmony with Albert Schweitzer's principle of reverence for life. With its stress on the "natural integrity of the places visited," the Audubon maxim also approaches Aldo Leopold's emphasis on the whole community in his land ethic. This last point also applies to the Antarctic Code's concerns with both the plants and animals of Antarctica to the extent of humans not even leaving footprints. In very fragile ecosystems, using Leopold's land ethic, the argument could be made that human visitors not be permitted to go to the site at all. Ecocentric holism would thwart any claims by human travelers to visit areas where their presence would compromise the stability and integrity of a particular ecosystem or even a specific plant or animal species.

Stanley Plog's maxim 1: "Protect what is natural and beautiful for the benefit of 'natives' and tourists," is based on anthropocentric principles. Protection is in service to the human community and there is no mention in this or in any of the other maxims Plog develops of inherent value in nature. Perhaps this should not be surprising since Plog's purpose in proposing the code in the first place is to develop "a code for leisure destination development to protect their futures as they grow over the years." Nature, in this view, is very much the resource or commodity that should be developed for human purposes and interests. The anthropocentric foundation is clear in Plog's maxims, although these maxims may be limited in serious ways as communities attempt to resolve the complex environmental issues raised by ecotourism. That being said, Plog's efforts are important for seeing the need for a code in the first place and for arguing for the importance of cooperation between environmental groups and the tourist industry. Nevertheless, how will this cooperation take place, especially given the fact that highly contentious issues are at stake? Frameworks for negotiating discourses need to be established and this task itself will not be easy to accomplish.

FRAMEWORKS FOR NEGOTIATING DISCOURSES ABOUT ECOTOURISM

Efforts to cooperate in solving problems in tourism need to be much more broadly based than Plog indicates. The niches of ecotourism and adventure travel provide an experimental ground for trying out ways to negotiate difficulties and differences among interested parties. Because ecotourism and adventure travel occur on a smaller scale than other types of tourism, experimenting in these areas may help provide insights for solving environmental problems relating to tourism on a broader scale. Furthermore, analyzing types of discourse that do occur in environmental conflicts and making suggestions for negotiating these discourses may offer ways to think about and to act in contentious areas like ecotourism and adventure travel.

There is much to be gained in examining examples of discourse and public hearings to analyze the content as well as the various communication methods used by participants. Such an analysis constitutes a chapter in Tarla Rai Peterson's Sharing the Earth. Although this analysis does not focus on issues of ecotourism, there are many similarities that one can imagine would play out in local debates on the impacts of ecotourism. In her chapter Peterson analyzes hearings that were conducted by Agriculture Canada in January 1990 on its plan to remove an entire bison herd from the Wood Buffalo National Park because of Brucellosis infection in the herd. The group conducting the hearings was called the Northern Diseased Bison Environmental Assessment Panel and it held seven hearings in various locations in northern Alberta and the Northwest Territories. Peterson shows that Agriculture Canada's role in conducting the hearings was open to serious question from the start since it had already submitted its plan for the removal of the entire herd and as such was an advocate for one of the proposed solutions. As it turned out, the Panel's final recommendations were only slightly different from what Agriculture Canada had proposed in the first place. Moreover, while the final report did acknowledge some of the objections forcefully made by aboriginal and environmental groups, the report did not respond to them in any serious way.

Peterson goes on to show, among other things, that certain types of discourse that she calls "technological discourse" used by the scientific experts and members of the government agencies were privileged over the "creative discourse" of the aboriginal and other local groups. The local people's long-term and close relationship with the buffalo herd was clearly not granted the same epistemological and policy value as that of the "experts" in various fields. These sharply contrasting communication methods resulted in conflicts and impasses throughout the hearings with neither side able to engage in a kind of meta-discourse that might have helped them negotiate their widely divergent assumptions, communications methods, and expectations during and after the hearings.

Without going into any greater detail with this important example and Peterson's insightful analysis, it suffices to note that these kinds of hearings and panels occur with great frequency as local communities, regional, and national jurisdictions and other stakeholders attempt to discuss and decide upon issues related to development, resources, and land uses. It is easy to imagine similar panels and commissions being established to deal with the possible impacts of ecotourism and adventure travel in far-flung and wilderness regions of the world. Could these discussions be constructed so that all participants would know that the information and knowledge they bring to the table would be fully heard, considered, and appropriated in meaningful ways? Much work would need to be done prior to such hearings and panels so that stakeholders could be assured that their concerns, interests, and communication methods would be fully honored. This is asking a lot, especially given unequal power relationships that exist, for example, between so-called "expert" and "non-expert" participants. With technical and scientific discourses given greater epistemological weight and often considered "objective" over against "subjective" local and lay discourses, the meta-discourses might turn out to be just as arduous and problematic as the hearings themselves. This is not to say that such meta-discourse should not be attempted. In fact, communication theorists along with proponents of discourse ethics may do a lot to help establish frameworks for negotiating such hearings that could be adopted by the participants who are entering into a forum to debate and decide a particular issue.

Two theoretical approaches may be fruitful in developing these frameworks: first, Habermas's discourse ethics and second, some versions of feminist ethics and political theory, in particular Seyla Benhabib's reflections on Hannah Arendt's work.

Jurgen Habermas's discourse ethics is proving to be one of the most promising developments in moral theory in the last few decades. That being said, it is also both contentious and hotly debated. For our purposes here, it is not so important to enter into these debates, especially the very vital one concerning the justification of the principle of universalizability (U) that stands as the central normative principle of Habermas's moral theory. According to this principle, a rational consensus on a proposed norm is reached and thus the norm is valid, if and only if:

> All affected can accept the consequences and the side effects its general observance can be anticipated to have for the satisfaction of everyone's interests (and these consequences are preferred to those of known alternative possibilities for regulation).

Again, apart from the many questions raised by the theory as a whole, what is valuable for issues and conflicts raised by ecotourism is the assertion that the interests of those actually affected by decisions are morally relevant and that moral validity depends on the real consensus of participants in actual discussions. Habermas's emphasis on intersub-jectivity adds a very important dimension to this moral theory and has particular relevance for the kinds of disputes that arise in the public arena, in this case, in questions about the moral aspects of ecotourism and its impacts. On the level of moral theory, much as been made of Habermas's efforts to construct a clear way to come to rational consensus among people who do not share the same history, traditions or value systems. This effort, both problematic and fruitful, has yet to be fully exploited and it would be most helpful to flesh this out, given the nature of debates surrounding public policy issues like land and resource management and in questions generated by ecotourism. As essential as these theoretical issues are, it is also in the actual debates and discussions among participants that other thorny conflicts arise. A kind of "application discourse" needs to be developed and drawing implications from Habermas's work would advance this project.

William Rehg has done much to draw out these implications in his work on Habermas in which Rehg argues that at the point of actual, real life discussions, participants need to express a kind of "rational trust" that the legal and political procedures put in place in the spirit of the principle of universalizability (U) will be adhered to and fulfilled. This notion of rational trust seems to contain its own criteria of rationality and have its own moral weight apart from the U principle:

> On this view, the individual's confidence in the validity of a decision is based not so much on his or her overview of the relevant arguments as on the procedures for processing various arguments and how faithfully their administrators carry them out.

Rehg's suggestions here go part of the way, but not far enough since he does not show how trust in the procedures can be considered rational nor does he fully elaborate on what these procedures may be. For instance, where would rituals, performance art, and storytelling fit in his definitions of rationality or how might they fit into legal and political procedures? Rehg imagines that on any particular issue many audiences may need to be involved, resulting, he claims, in dispersed consensus. Such a consensus, freely and fully arrived at by all participants, needs to be seen not merely as a compromise (i.e., the best one can hope for under the real circumstance of the debate), but as a genuine, moral decision that can be accepted by all as a morally, and not only legally or procedurally, binding. This is a tall order in the hurly-burly, real life conditions of such debates.

Constructing these spaces requires that all participants cultivate an "enlarged mentality" in which all learn how to reason, understand and appreciate the standpoint of other participants. For Arendt and then Benhabib, such understanding is the outcome of political meeting and speaking in the public space. The multiple perspectives that constitute the political are only revealed in the speaking acts of those who are willing to engage in the public drama in the first place. Public space is thereby created by such collective and contested actions. Participants in the public space take the standpoint of others and exercise what Arendt called "transcending judgment" by which one accords each person the moral respect to consider seriously the other's point of view. This does not mean agreeing with or assuming the other standpoint by foregoing one's own; rather the challenge is to think from the other's standpoint and to listen in a genuinely open way to what the other's standpoint entails. When this happens, the associational public space is enacted in which most likely conflict occurs but without competition. Storytelling, ritual, memories, and myths can all contribute to the creation of associational public space in which knowledge and testimony are expanded to include narratives of all sorts.

Discourse in this setting is not judged solely on principles of rationality or objectivity, but it depends on the relational processes that are enacted within the associational public space. It is important to note that these relational processes do not presuppose the existence of a robust political space, but in fact lead to its creation through the enactment of the multiple discourses in the events themselves. In other words, the end product or the final decisions are not already set, as they were in the Northern Diseased

Bison Environmental Assessment Panel hearings, but would emerge as a result of the hearings themselves. In fact, Agriculture Canada's recommendation to remove the entire bison herd was made before the hearings actually took place. In the new framework presented here, both the technological and creative discourses would be accorded parity within the context of the enlarged mentality of all participants who are committed to creating an associational public space. Negotiating discourses in this associational public space could be a more fruitful enterprise than the agonistic and competitive models currently in use. Conflicts centering on ecotourism and adventure travel would constitute interesting cases in these associational public spaces that also used norms generated from discourse ethics and feminist ethical theories.

In addition to the principles delineated above from Habermas's discourse ethics and from Benhabib and Arendt's works, actual discussions on ecotourism and adventure travel should include the following features:

- As has already occurred in many debates-discussions need to be multi-disciplinary in approach, including but not limited to the tourist industry, environmental organizations, environmental studies, government agencies, representatives of local populations, ethicists, and public policy makers.
- All factors impinging upon a particular tourist destination need to be considered: economic, environmental (in the most complete sense), social, political, ethical, and aesthetic. An agreed upon method of environmental accounting needs to be devised. All stakeholders participate in debates, discussions, and decision-making. Working groups accommodate "expert" and "non-expert" discourse and ways of thinking. Non-linear discourse is accorded due weight in discussion and decisionmaking, e.g., traditional myths, storytelling, rituals, and religious beliefs are included in significant ways and not simply as tokenism.
- Advocacies are delineated, i.e., who speaks for non-rational, non-verbal and inanimate parts of the community. This is addressed and resolved before discussions take place. Indigenous peoples with claims to specific sites have the right to participate or not in such discussions and decision-making. They may refuse to participate and withhold their territory and specific sites from any development or contacts they consider unsuitable and inappropriate. As Barry Lopez remarked in a

talk he gave recently at the New York City Public Library, we may invite the local people to meet us at the trailhead. They may choose to accept or reject the invitation.

- Models for conducting discussion and decision-making are decided upon before discussion of issues occurs. Facilitators, discussion leaders and group process consultants who have no stake in the issues are given the task to implement and monitor the discussions and decision-making.

Parts of these frameworks are already being enacted. Nature organizations, ecotourism operators, indigenous peoples, public officials and others are creating and participating in the process of formulating agreements and codes for responsible tourism. There is no doubt that working to create associational public spaces and working within these frameworks will be difficult and time-consuming. It will also require good will, openness, and honesty on all sides.

As is clear from the analysis so far, not only are there many serious issues to address in the substance of ecotourism, but also in the processes themselves that need to be created to address them. The discourse ethics of Habermas augmented by the feminist political and ethical theories of Benhabib and Arendt are fruitful theoretical approaches that also have the room to deal with policy and practical applications as these debates and discourses get underway.

Another important point is the need for greater environmental education and public awareness about both the pleasures and ecological perils of tourism. Web sites and the Internet (list serves, chat rooms, etc.) are excellent tools for education in this broad sense. With our students, we should carefully combine research, field work, and media of all sorts to help them become environmentally responsible global citizens.

With regard to ecotourism and adventure travel, we propose the formation of councils or commissions that would grant a kind of "certification" or "seal of approval" that ecotour operators may apply for to use in their promotion and marketing. Then travelers may decide whether or not they will patronize ecotours based on the environmental track records of the tour operators. Such certification could also be made available to ecotour destinations. Travelers would then have the confidence that the local authorities in particular locations are doing everything in their power to maintain and enhance the complete ecosystems in their jurisdictions.

There are other courses of action open to interested parties. National and international environmental groups lobby parliaments and legislatures for better environmental regulations and enforcement. These are daunting and difficult tasks, but such legal improvements are needed for long-term solutions. International treaties and trade agreements also need to consider the short and long-range environmental consequences of tourism in negotiations and regulations.

References

Barkin, D. (2002). *Ecotourism for sustainable regional development.* Current Issues in Tourism. pp. 5(3–4):245–253.

Batta R.N.(2000). *Tourism And The Environment: A Quest for Sustainability.* Indus Publishing Company, New Delhi. pp. 203.

Boo, E. (1990). *Ecotourism: The potentials and pitfalls.* Volumes 1 and 2. Washington D.C.: World Wildlife Fund.

Brandon, K. (1996). *Ecotourism and conservation: A review of key issues.* World Bank Environmental Department Paper No. 033. Washington D.C.: The World Bank.

Ceballos-Lascurain, H. (1996). *Tourism, Ecotourism and Protected Areas.* Gland, Switzerland: IUCN.

Holden, A. (2000). *Environment and Tourism.* London: Routledge.

2

Global Code of Ethics for Tourism

The Global Code of Ethics for Tourism (GCET) is a comprehensive set of principles whose purpose is to guide stakeholders in tourism development: central and local governments, local communities, the tourism industry and its professionals, as well as visitors, both international and domestic.

The Code was called for in a resolution of the WTO General Assembly meeting in Istanbul in 1997. Over the following two years, a special committee for the preparation of the Global Code of Ethics was formed and a draft document was prepared by the Secretary-General and the legal adviser to WTO in consultation with WTO Business Council, WTO's Regional Commissions, and the WTO Executive Council.

The United Nations Commission on Sustainable Development meeting in New York in April, 1999 endorsed the concept of the Code and requested WTO to seek further input from the private sector, non-governmental organizations and labour organizations. Written comments on the code were received from more than 70 WTO Member States and other entities. The resulting 10 point Global Code of Ethics for Tourism — the culmination of an extensive consultative process- was approved unanimously by the WTO General Assembly meeting in Santiago in October 1999.

The United Nations Economic and Social Council (ECOSOC), in its substantive session of July 2001, adopted a draft resolution on the Code of Ethics and called on the UN General Assembly to give recognition to the Code. The official recognition by the UN General Assembly to the Global

Code of Ethics for Tourism came on 21 December 2001, through its resolution A/RES/56/212, by which it further encouraged the World Tourism Organization to promote an effective follow-up of the Code.

Article 1. Tourism's Contribution to Mutual Understanding and Respect between Peoples and Societies

The understanding and promotion of the ethical values common to humanity, with an attitude of tolerance and respect for the diversity of religious, philosophical and moral beliefs, are both the foundation and the consequence of responsible tourism; stakeholders in tourism development and tourists themselves should observe the social and cultural traditions and practices of all peoples, including those of minorities and indigenous peoples and to recognise their worth;

Tourism activities should be conducted in harmony with the attributes and traditions of the host regions and countries and in respect for their laws, practices and customs;

The host communities, on the one hand, and local professionals, on the other, should acquaint themselves with and respect the tourists who visit them and find out about their lifestyles, tastes and expectations; the education and training imparted to professionals contribute to a hospitable welcome;

It is the task of the public authorities to provide protection for tourists and visitors and their belongings; they must pay particular attention to the safety of foreign tourists owing to the particular vulnerability they may have; they should facilitate the introduction of specific means of information, prevention, security, insurance and assistance consistent with their needs; any attacks, assaults, kidnappings or threats against tourists or workers in the tourism industry, as well as the wilful destruction of tourism facilities or of elements of cultural or natural heritage should be severely condemned and punished in accordance with their respective national laws;

When travelling, tourists and visitors should not commit any criminal act or any act considered criminal by the laws of the country visited and abstain from any conduct felt to be offensive or injurious by the local populations, or likely to damage the local environment; they should refrain from all trafficking in illicit drugs, arms, antiques, protected species and products and substances that are dangerous or prohibited by national regulations;

Tourists and visitors have the responsibility to acquaint themselves, even before their departure, with the characteristics of the countries they are preparing to visit; they must be aware of the health and security risks inherent in any travel outside their usual environment and behave in such a way as to minimise those risks;

Article 2. Tourism as a Vehicle for Individual and Collective Fulfilment

Tourism, the activity most frequently associated with rest and relaxation, sport and access to culture and nature, should be planned and practised as a privileged means of individual and collective fulfilment; when practised with a sufficiently open mind, it is an irreplaceable factor of self-education, mutual tolerance and for learning about the legitimate differences between peoples and cultures and their diversity;

Tourism activities should respect the equality of men and women; they should promote human rights and, more particularly, the individual rights of the most vulnerable groups, notably children, the elderly, the handicapped, ethnic minorities and indigenous peoples;

The exploitation of human beings in any form, particularly sexual, especially when applied to children, conflicts with the fundamental aims of tourism and is the negation of tourism; as such, in accordance with international law, it should be energetically combatted with the cooperation of all the States concerned and penalised without concession by the national legislation of both the countries visited and the countries of the perpetrators of these acts, even when they are carried out abroad;

Travel for purposes of religion, health, education and cultural or linguistic exchanges are particularly beneficial forms of tourism, which deserve encouragement;

The introduction into curricula of education about the value of tourist exchanges, their economic, social and cultural benefits, and also their risks, should be encouraged;

Article 3. Tourism, a Factor of Sustainable Development

All the stakeholders in tourism development should safeguard the natural environment with a view to achieving sound, continuous and sustainable economic growth geared to satisfying equitably the needs and aspirations of present and future generations;

All forms of tourism development that are conducive to saving rare and precious resources, in particular water and energy, as well as avoiding so far as possible waste production, should be given priority and encouraged by national, regional and local public authorities;

The staggering in time and space of tourist and visitor flows, particularly those resulting from paid leave and school holidays, and a more even distribution of holidays should be sought so as to reduce the pressure of tourism activity on the environment and enhance its beneficial impact on the tourism industry and the local economy;

Tourism infrastructure should be designed and tourism activities programmed in such a way as to protect the natural heritage composed of ecosystems and biodiversity and to preserve endangered species of wildlife; the stakeholders in tourism development, and especially professionals, should agree to the imposition of limitations or constraints on their activities when these are exercised in particularly sensitive areas: desert, polar or high mountain regions, coastal areas, tropical forests or wetlands, propitious to the creation of nature reserves or protected areas;

Nature tourism and ecotourism are recognised as being particularly conducive to enriching and enhancing the standing of tourism, provided they respect the natural heritage and local populations and are in keeping with the carrying capacity of the sites;

Article 4. Tourism, a User of the Cultural Heritage of Mankind and Contributor to its Enhancement

Tourism resources belong to the common heritage of mankind; the communities in whose territories they are situated have particular rights and obligations to them;

Tourism policies and activities should be conducted with respect for the artistic, archaeological and cultural heritage, which they should protect and pass on to future generations; particular care should be devoted to preserving and upgrading monuments, shrines and museums as well as archaeological and historic sites which must be widely open to tourist visits; encouragement should be given to public access to privately-owned cultural property and monuments, with respect for the rights of their owners, as well as to religious buildings, without prejudice to normal needs of worship;

Financial resources derived from visits to cultural sites and monuments should, at least in part, be used for the upkeep, safeguard, development and embellishment of this heritage;

Tourism activity should be planned in such a way as to allow traditional cultural products, crafts and folklore to survive and flourish, rather than causing them to degenerate and become standardised;

Article 5. Tourism, a Beneficial Activity for Host Countries and Communities

Local populations should be associated with tourism activities and share equitably in the economic, social and cultural benefits they generate, and particularly in the creation of direct and indirect jobs resulting from them;

Tourism policies should be applied in such a way as to help to raise the standard of living of the populations of the regions visited and meet their needs; the planning and architectural approach to and operation of tourism resorts and accommodation should aim to integrate them, to the extent possible, in the local economic and social fabric; where skills are equal, priority should be given to local manpower;

Special attention should be paid to the specific problems of coastal areas and island territories and to vulnerable rural or mountain regions, for which tourism often represents a rare opportunity for development in the face of the decline of traditional economic activities;

Tourism professionals, particularly investors, governed by the regulations laid down by the public authorities, should carry out studies of the impact of their development projects on the environment and natural surroundings; they should also deliver, with the greatest transparency and objectivity, information on their future programmes and their foreseeable repercussions and foster dialogue on their contents with the populations concerned;

Article 6. Obligations of Stakeholders in Tourism Development

Tourism professionals have an obligation to provide tourists with objective and honest information on their places of destination and on the conditions of travel, hospitality and stays; they should ensure that the contractual clauses proposed to their customers are readily understandable as to the nature, price and quality of the services they commit themselves to providing and the financial compensation payable by them in the event of a unilateral breach of contract on their part;

Tourism professionals, insofar as it depends on them, should show concern, in co-operation with the public authorities, for the security and safety, accident prevention, health protection and food safety of those who seek their services; likewise, they should ensure the existence of suitable systems of insurance and assistance; they should accept the reporting obligations prescribed by national regulations and pay fair compensation in the event of failure to observe their contractual obligations

Tourism professionals, so far as this depends on them, should contribute to the cultural and spiritual fulfilment of tourists and allow them, during their travels, to practise their religions;

The public authorities of the generating States and the host countries, in cooperation with the professionals concerned and their associations, should ensure that the necessary mechanisms are in place for the repatriation of tourists in the event of the bankruptcy of the enterprise that organised their travel;

Governments have the right – and the duty - especially in a crisis, to inform their nationals of the difficult circumstances, or even the dangers they may encounter during their travels abroad; it is their responsibility however to issue such information without prejudicing in an unjustified or exaggerated manner the tourism industry of the host countries and the interests of their own operators; the contents of travel advisories should therefore be discussed beforehand with the authorities of the host countries and the professionals concerned; recommendations formulated should be strictly proportionate to the gravity of the situations encountered and confined to the geographical areas where the insecurity has arisen; such advisories should be qualified or cancelled as soon as a return to normality permits;

The press, and particularly the specialised travel press and the other media, including modern means of electronic communication, should issue honest and balanced information on events and situations that could influence the flow of tourists; they should also provide accurate and reliable information to the consumers of tourism services; the new communication and electronic commerce technologies should also be developed and used for this purpose; as is the case for the media, they should not in any way promote sex tourism;

Article 7. Right to Tourism

The prospect of direct and personal access to the discovery and enjoyment of the planet's resources constitutes a right equally open to all the world's inhabitants; the increasingly extensive participation in national and international tourism should be regarded as one of the best possible expressions of the sustained growth of free time, and obstacles should not be placed in its way;

The universal right to tourism must be regarded as the corollary of the right to rest and leisure, including reasonable limitation of working hours and periodic holidays with pay, guaranteed by Article 24 of the Universal Declaration of Human Rights and Article 7.d of the International Covenant on Economic, Social and Cultural Rights;

Social tourism, and in particular associative tourism, which facilitates widespread access to leisure, travel and holidays, should be developed with the support of the public authorities;

Family, youth, student and senior tourism and tourism for people with disabilities, should be encouraged and facilitated;

Article 8. Liberty of Tourist Movements

Tourists and visitors should benefit, in compliance with international law and national legislation, from the liberty to move within their countries and from one State to another, in accordance with Article 13 of the Universal Declaration of Human Rights; they should have access to places of transit and stay and to tourism and cultural sites without being subject to excessive formalities or discrimination;

Tourists and visitors should have access to all available forms of communication, internal or external; they should benefit from prompt and easy access to local administrative, legal and health services; they should be free to contact the consular representatives of their countries of origin in compliance with the diplomatic conventions in force;

Tourists and visitors should benefit from the same rights as the citizens of the country visited concerning the confidentiality of the personal data and information concerning them, especially when these are stored electronically;

Administrative procedures relating to border crossings whether they fall within the competence of States or result from international agreements, such as visas or health and customs formalities, should be adapted, so far as possible, so as to facilitate to the maximum freedom of travel and widespread access to international tourism; agreements between groups of countries to harmonise and simplify these procedures should be encouraged; specific taxes and levies penalising the tourism industry and undermining its competitiveness should be gradually phased out or corrected;

So far as the economic situation of the countries from which they come permits, travellers should have access to allowances of convertible currencies needed for their travels;

Article 9. Rights of the Workers and Entrepreneurs in the Tourism Industry

The fundamental rights of salaried and self-employed workers in the tourism industry and related activities, should be guaranteed under the supervision of the national and local administrations, both of their States of origin and of the host countries with particular care, given the specific constraints linked in particular to the seasonality of their activity, the global dimension of their industry and the flexibility often required of them by the nature of their work;

Salaried and self-employed workers in the tourism industry and related activities have the right and the duty to acquire appropriate initial and continuous training; they should be given adequate social protection; job insecurity should be limited so far as possible; and a specific status, with particular regard to their social welfare, should be offered to seasonal workers in the sector;

Any natural or legal person, provided he, she or it has the necessary abilities and skills, should be entitled to develop a professional activity in the field of tourism under existing national laws; entrepreneurs and investors - especially in the area of small and medium-sized enterprises - should be entitled to free access to the tourism sector with a minimum of legal or administrative restrictions;

Exchanges of experience offered to executives and workers, whether salaried or not, from different countries, contributes to foster the development of the world tourism industry; these movements should be facilitated so far as possible in compliance with the applicable national laws and international conventions;

As an irreplaceable factor of solidarity in the development and dynamic growth of international exchanges, multinational enterprises of the tourism industry should not exploit the dominant positions they sometimes occupy; they should avoid becoming the vehicles of cultural and social models artificially imposed on the host communities; in exchange for their freedom to invest and trade which should be fully recognised, they should involve themselves in local development, avoiding, by the excessive repatriation of their profits or their induced imports, a reduction of their contribution to the economies in which they are established;

Partnership and the establishment of balanced relations between enterprises of generating and receiving countries contribute to the sustainable development of tourism and an equitable distribution of the benefits of its growth;

Article 10. Implementation of the Principles of the Global Code of Ethics for Tourism

The public and private stakeholders in tourism development should cooperate in the implementation of these principles and monitor their effective application;

The stakeholders in tourism development should recognise the role of international institutions, among which the World Tourism Organisation ranks first, and non-governmental organisations with competence in the field of tourism promotion and development, the protection of human rights, the environment or health, with due respect for the general principles of international law;

The same stakeholders should demonstrate their intention to refer any disputes concerning the application or interpretation of the Global Code of Ethics for Tourism for conciliation to an impartial third body known as the World Committee on Tourism Ethics.

3

Québec Declaration on Ecotourism

In the framework of the UN International Year of Ecotourism, 2002, under the aegis of the United Nations Environment Programme (UNEP) and the World Tourism Organisation (WTO), over one thousand participants coming from 132 countries, from the public, private and non-governmental sectors met at the World Ecotourism Summit, hosted in Quebec City, Canada, by Tourisme Quebec and the Canadian Tourism Commission, between 19 and 22 May 2002.

The Quebec Summit represented the culmination of 18 preparatory meetings held in 2001 and 2002, involving over 3,000 representatives from national and local governments including the tourism, environment and other administrations, private ecotourism businesses and their trade associations, non-governmental organisations, academic institutions and consultants, intergovernmental organisations, and indigenous and local communities.

This document takes into account the preparatory process, as well as the discussions held during the Summit. It is the result of a multistakeholder dialogue, although it is not a negotiated document. Its main purpose is the setting of a preliminary agenda and a set of recommendations for the development of ecotourism activities in the context of sustainable development.

The participants at the Summit acknowledge the World Summit on Sustainable Development (WSSD) in Johannesburg, August/September 2002, as the ground-setting event for international policy in the next 10 years, and emphasise that, as a leading industry, the sustainability of tourism should be a priority at WSSD due to its potential contribution to poverty alleviation and environmental protection in endangered ecosystems. Participants therefore request the UN, its organisations and member governments represented at this Summit to disseminate the following Declaration and other results from the World Ecotourism Summit at the WSSD.

The participants to the World Ecotourism Summit, aware of the limitations of this consultative process to incorporate the input of the large variety of ecotourism stakeholders, particularly non-governmental organisations (NGOs) and local and indigenous communities,

Recognise that ecotourism embraces the principles of sustainable tourism, concerning the economic, social and environmental impacts of tourism. It also embraces the following specific principles which distinguish it from the wider concept of sustainable tourism:

— Contributes actively to the conservation of natural and cultural heritage,

— Includes local and indigenous communities in its planning, development and operation, and contributing to their well-being,

— Interprets the natural and cultural heritage of the destination to visitors,

— Lends itself better to independent travellers, as well as to organised tours for small size groups.

Acknowledge that tourism has significant and complex social, economic and environmental implications, which can bring both benefits and costs to the environment and local communities,

Consider the growing interest of people in travelling to natural areas, both on land and sea,

Recognise that ecotourism has provided a leadership role in introducing sustainability practices to the tourism sector,

Emphasise that ecotourism should continue to contribute to make the overall tourism industry more sustainable, by increasing economic and social benefits for host communities, actively contributing to the conservation of natural resources and the cultural integrity of host communities, and by increasing awareness of all travellers towards the conservation of natural and cultural heritage,

Recognise the cultural diversity associated with many natural areas, particularly because of the historical presence of local and indigenous communities, of which some have maintained their traditional knowledge, uses and practices many of which have proven to be sustainable over the centuries,

Reiterate that funding for the conservation and management of biodiverse and culturally rich protected areas has been documented to be inadequate worldwide,

Recognise further that many of these areas are home to peoples often living in poverty, who frequently lack adequate health care, education facilities, communications systems, and other infrastructure required for genuine development opportunity,

Affirm that different forms of tourism, especially ecotourism, if managed in a sustainable manner can represent a valuable economic opportunity for local and indigenous populations and their cultures and for the conservation and sustainable use of nature for future generations and can be a leading source of revenues for protected areas,

Emphasise that at the same time, wherever and whenever tourism in natural and rural areas is not properly planned, developed and managed, it contributes to the deterioration of natural landscapes, threats to wildlife and biodiversity, marine and coastal pollution, poor water quality, poverty, displacement of indigenous and local communities, and the erosion of cultural traditions,

Acknowledge that ecotourism development must consider and respect the land and property rights, and, where recognised, the right to self-determination and cultural sovereignty of indigenous and local communities, including their protected, sensitive and sacred sites as well as their traditional knowledge,

Stress that to achieve equitable social, economic and environmental benefits from ecotourism and other forms of tourism in natural areas, and to minimise or avoid potential negative impacts, participative planning mechanisms are needed that allow local and indigenous communities, in a transparent way, to define and regulate the use of their areas at the local level, including the right to opt out of tourism development,

Understand that small and micro businesses seeking to meet social and environmental objectives are key partners in ecotourism and are often operating in a development climate that does not provide suitable financial and marketing support for ecotourism,

Recognise that to improve the chances of survival of small-, medium-, and micro enterprises further understanding of the ecotourism market will be required through market research, specialised credit instruments for tourism businesses, grants for external costs, incentives for the use of sustainable energy and innovative technical solutions, and an emphasis on developing skills not only in business but within government and those seeking to support business solutions,

Accept the need to avoid discrimination between people, whether by race, gender or other personal circumstances, with respect to their involvement in ecotourism as consumers or suppliers,

Recognise that visitors have a responsibility to the sustainability of the destination and the global environment through their travel choice, behaviour and activities, and that therefore it is important to communicate to them the qualities and sensitivities of destinations,

In light of the above, the participants to the World Ecotourism Summit, having met in Quebec City, from 19 to 22 May 2002, produced a series of recommendations, which they propose to governments, the private sector, non-governmental organisations, community-based associations, academic and research institutions, inter-governmental organisations, international financial institutions, development assistance agencies, and indigenous and local communities, as follows:

A. To National, Regional and Local Governments

1. Formulate national, regional and local ecotourism policies and development strategies that are consistent with the overall objectives of sustainable development, and to do so through a wide consultation process with those who are likely to become involved in, affect, or be affected by ecotourism activities;
2. Guarantee -in conjunction with local and indigenous communities, the private sector, NGOs and all ecotourism stakeholders- the protection of nature, local and indigenous cultures and specially traditional knowledge, genetic resources, rights to land and property, as well as rights to water;
3. Ensure the involvement, appropriate participation and necessary coordination of all the relevant public institutions at the national, provincial and local level, (including the establishment of inter-

ministerial working groups as appropriate) at different stages in the ecotourism process, while at the same time opening and facilitating the participation of other stakeholders in ecotourism-related decisions. Furthermore, adequate budgetary mechanisms and appropriate legislative frameworks need to be set up to allow implementation of the objectives and goals set up by these multistakeholder bodies;

4. Include in the above framework the necessary regulatory and monitoring mechanisms at the national, regional and local levels, including objective sustainability indicators jointly agreed with all stakeholders and environmental impact assessment studies to be used as feedback mechanism. Results of monitoring should be made available to the general public;
5. Develop regulatory mechanisms for internalisation of environmental costs in all aspects of the tourism product, including international transport;
6. Develop the local and municipal capacity to implement growth management tools such as zoning, and participatory land-use planning not only in protected areas but in buffer zones and other ecotourism development zones;
7. Use internationally approved and reviewed guidelines to develop certification schemes, ecolabels and other voluntary initiatives geared towards sustainability in ecotourism, encouraging private operators to join such schemes and promoting their recognition by consumers. However, certification systems should reflect regional and local criteria. Build capacity and provide financial support to make these schemes accessible to small and medium enterprises (SMEs). In addition, monitoring and a regulatory framework are necessary to support effective implementation of these schemes;
8. Ensure the provision of technical, financial and human resources development support to micro, small and medium-sized firms, which are the core of ecotourism, with a view to enable them to start, grow and develop their businesses in a sustainable manner;
9. Define appropriate policies, management plans, and interpretation programmes for visitors, and earmark adequate sources of funding for natural areas to manage visitor numbers, protect vulnerable ecosystems, and the sustainable use of sensitive habitats. Such plans should include

clear norms, direct and indirect management strategies, and regulations with the funds to ensure monitoring of social and environmental impacts for all ecotourism businesses operating in the area, as well as for tourists wishing to visit them;

10. Include micro, small and medium-sized ecotourism companies, as well as communitybased and NGO-based ecotourism operations in the overall promotional strategies and programmes carried out by the National Tourism Administration, both in the international and domestic markets;
11. Encourage and support the creation of regional networks and cooperation for promotion and marketing of ecotourism products at the international and national levels;
12. Provide incentives to tourism operators and other service providers (such as marketing and promotion advantages) for them to adopt ecotourism principles and make their operations more environmentally, socially and culturally responsible;
13. Ensure that basic environmental and health standards are identified and met by all ecotourism development even in the most rural areas. This should include aspects such as site selection, planning, design, the treatment of solid waste, sewage, and the protection of watersheds, etc., and ensure also that ecotourism development strategies are not undertaken by governments without investment in sustainable infrastructure and the reinforcement of local/municipal capabilities to regulate and monitor such aspects;
14. Institute baseline environmental impact assessment (EIA) studies and surveys that record the social environmental state of destinations, with special attention to endangered species, and invest, or support institutions that invest in research programmes on ecotourism and sustainable tourism;
15. Support the further implementation of the international principles, guidelines and codes of ethics for sustainable tourism (e.g. such as those proposed by UNEP, WTO, the Convention on Biological Diversity, the UN Commission on Sustainable Development and the International Labor Organisation) for the enhancement of international and national legal frameworks, policies and master plans to implement the concept of sustainable development into tourism;

16. Consider as one option the reallocation of tenure and management of public lands, from extractive or intensive productive sectors to tourism combined with conservation, wherever this is likely to improve the net social, economic and environmental benefit for the community concerned;
17. Promote and develop educational programmes addressed to children and young people to enhance awareness about nature conservation and sustainable use, local and indigenous cultures and their relationship with ecotourism;
18. Promote collaboration between outbound tour operators and incoming operators and other service providers and NGOs at the destination to further educate tourists and influence their behaviour at destinations, especially those in developing countries;
19. Incorporate sustainable transportation principles in the planning and design of access and transportation systems, and encourage tour operators and the travelling public to make soft mobility choices.

B. To the Private Sector

20. Bear in mind that for ecotourism businesses to be sustainable, they need to be profitable for all stakeholders involved, including the projects' owners, investors, managers and employees, as well as the communities and the conservation organisations of natural areas where it takes place;
21. Conceive, develop and conduct their businesses minimising negative effects on, and positively contributing to, the conservation of sensitive ecosystems and the environment in general, and directly benefiting and including local and indigenous communities;
22. Ensure that the design, planning, development and operation of ecotourism facilities incorporates sustainability principles, such as sensitive site design and community sense of place, as well as conservation of water, energy and materials, and accessibility to all categories of population without discrimination;
23. Adopt as appropriate a reliable certification or other systems of voluntary regulation, such as ecolabels, in order to demonstrate to their potential clients their adherence to sustainability principles and the soundness of the products and services they offer;

24. Cooperate with governmental and non-governmental organisations in charge of protected natural areas and conservation of biodiversity, ensuring that ecotourism operations are practised according to the management plans and other regulations prevailing in those areas, so as to minimise any negative impacts upon them while enhancing the quality of the tourism experience and contribute financially to the conservation of natural resources;
25. Make increasing use of local materials and products, as well as local logistical and human resource inputs in their operations, in order to maintain the overall authenticity of the ecotourism product and increase the proportion of financial and other benefits that remain at the destination. To achieve this, private operators should invest in the training of the local workforce;
26. Ensure that the supply chain used in building up an ecotourism operation is thoroughly sustainable and consistent with the level of sustainability aimed at in the final product or service to be offered to the customer;
27. Work actively with indigenous leadership and local communities to ensure that indigenous cultures and communities are depicted accurately and with respect, and that their staff and guests are well and accurately informed regarding local and indigenous sites, customs and history;
28. Promote among their clients an ethical and environmentally conscious behaviour vis-àvis the ecotourism destinations visited, such as by environmental education or by encouraging voluntary contributions to support local community or conservation initiatives;
29. Generate awareness among all management and staff of local, national and global environmental and cultural issues through ongoing environmental education, and support the contribution that they and their families can make to conservation, community economic development and poverty alleviation;
30. Diversify their offer by developing a wide range of tourist activities at a given destination and by extending their operations to different destinations in order to spread the potential benefits of ecotourism and to avoid overcrowding some selected ecotourism sites, thus threatening their long-term sustainability. In this regard, private operators are urged

to respect, and contribute to, established visitor impact management systems of ecotourism destinations;

31. Create and develop funding mechanisms for the operation of business associations or cooperatives that can assist with ecotourism training, marketing, product development, research and financing;
32. Ensure an equitable distribution of financial benefits from ecotourism revenues between international, outbound and incoming tour operators, local service providers and local communities through appropriate instruments and strategic alliances;
33. Formulate and implement company policies for sustainability with a view to applying them in each part of their operations.

C. To Non-governmental Organisations, Community-based Associations, Academic and Research Institutions

34. Provide technical, financial, educational, capacity building and other support to ecotourism destinations, host community organisations, small businesses and the corresponding local authorities in order to ensure that appropriate policies, development and management guidelines, and monitoring mechanisms are being applied towards sustainability;
35. Monitor and conduct research on the actual impacts of ecotourism activities upon ecosystems, biodiversity, local and indigenous cultures and the socio-economic fabric of the ecotourism destinations;
36. Cooperate with public and private organisations ensuring that the data and information generated through research is channeled to support decision-making processes in ecotourism development and management;
37. Cooperate with research institutions to develop the most adequate and practical solutions to ecotourism development issues.

D. To Inter-governmental Organisations, International Financial Institutions and Development Assistance Agencies

38. Develop and assist in the implementation of national and local policy and planning guidelines and evaluation frameworks for ecotourism and its relationships with biodiversity conservation, socio-economic development, respect of human rights, poverty alleviation, nature conservation and other objectives of sustainable development, and to

intensify the transfer of such know-how to all countries. Special attention should be paid to countries in a developing stage or least developed status, to small island developing States and to countries with mountain areas, considering that 2002 is also designated as the International Year of Mountains by the UN;

39. Build capacity for regional, national and local organisations for the formulation and application of ecotourism policies and plans, based on international guidelines;
40. Develop or adopt, as appropriate, international standards and financial mechanisms for ecotourism certification systems that take into account the needs of small and medium enterprises and facilitates their access to those procedures, and support their implementation;
41. Incorporate multistakeholder dialogue processes into policies, guidelines and projects at the global, regional and national levels for the exchange of experiences between countries and sectors involved in ecotourism;
42. Strengthen efforts in identifying the factors that determine the success or failure of ecotourism ventures throughout the world, in order to transfer such experiences and best practices to other nations, by means of publications, field missions, training seminars and technical assistance projects; UNEP, WTO and other international organisations should continue and expand the international dialogue after the Summit on sustainable tourism and ecotourism issues, for example by conducting periodical reviews of ecotourism development through international and regional forums;
43. Adapt as necessary their financial facilities and lending conditions and procedures to suit the needs of micro-, small- and medium-sized ecotourism firms that are the core of this industry, as a condition to ensure its long term economic sustainability;
44. Develop the internal human resource capacity to support sustainable tourism and ecotourism as a development sub-sector in itself and to ensure that internal expertise, research, and documentation are in place to oversee the use of ecotourism as a sustainable development tool;
45. Develop financial mechanisms for training and capacity building, that takes into account the time and resources required to successfully enable local communities and indigenous peoples to participate equitably in ecotourism development.

E. To Local and Indigenous Communities

In addition to all the references to local and indigenous communities made in the preceding paragraphs of this Declaration, (in particular para. 5, 8, 9 and 10 on page 2; para. 1 on page 3; in A 2 and 17; B 21 and 27; C 35; D 45) participants addressed the following recommendations to the local and indigenous communities themselves:

46. As part of a community vision for development, that may include ecotourism, define and implement a strategy for improving collective benefits for the community through ecotourism development including human, physical, financial, and social capital development, and improved access to technical information;
47. Strengthen, nurture and encourage the community's ability to maintain and use traditional skills, particularly home-based arts and crafts, agricultural produce, traditional housing and landscaping that use local natural resources in a sustainable manner.

F. To the World Summit on Sustainable Development (WSSD)

48. Recognise the need to apply the principles of sustainable development to tourism, and the exemplary role of ecotourism in generating economic, social and environmental benefits;
49. Integrate the role of tourism, including ecotourism, in the outcomes expected at WSSD.

4

Ecotourism: Planning and Management

Ecotourism is a relatively new concept, and it is still often misunderstood or misused. Some people have abused the term to attract conservation conscious travelers to what, in reality, are simply nature tourism programs which may cause negative environmental and social impacts. While the term was first heard in the 1980s, the first broadly accepted definition, and one which continues to be a valid "nutshell" definition was established by The (International) Ecotourism Society in 1990:

> Responsible travel to natural areas that conserves the environment and improves the well-being of local people.

As awareness and experience of the activity has grown, so has our need for a more comprehensive and detailed definition. Most recently (1999), Martha Honey has proposed an excellent, more detailed version:

> Ecotourism is travel to fragile, pristine and usually protected areas that strives to be low impact and (usually) small scale. It helps educate the traveler; provides funds for conservation; directly benefits the economic development and political empowerment of local communities; and fosters respect for different cultures and for human rights.

However, consensus exists among organizations involved with ecotourism (including The Nature Conservancy) around the definition adopted in 1996 by the World Conservation Union (IUCN) which describes ecotourism as:

> Environmentally responsible travel and visitation to natural areas, in order to enjoy and appreciate nature (and any accompanying cultural features, both past and present) that promote conservation, have a low visitor impact and provide for beneficially active socio-economic involvement of local peoples.

The Nature Conservancy has adopted the concept of ecotourism as the type of tourism that it recommends its partners use in most protected area management, especially for national parks and other areas with fairly strict conservation objectives. For The Nature Conservancy, ecotourism represents an excellent means for benefiting both local people and the protected area in question. It is an ideal component of a sustainable development strategy where natural resources can be utilized as tourism attractions without causing harm to the natural area. An important tool for protected area management and development, ecotourism must be implemented in a flexible manner.

Ecotourism Participants

Ahuge range of players with varying interests and goals participates in ecotourism. Some play more prominent roles than others, but almost all are represented in the development and management of ecotourism sites. A key to the success of ecotourism is the formation of strong partnerships so that the multiple goals of conservation and equitable development can be met. Partnerships may be difficult because of the number of players involved and their different needs, but forging relationships is essential. The key players can be classified as: protected area personnel, community organizations and individuals, private sector tourism industry members and a variety of government officials and nongovernmental organizations. Their effective interaction creates effective ecotourism.

Core Decision Makers

Protected Area Managers

Ecotourism involving protected areas places those in charge of the areas in a challenging position. Protected area personnel are often biologists, botanists or wildlife specialists whose job is to protect significant marine and terrestrial sites. Their key duties usually involve conducting inventories, managing wildlife populations and maintaining visitor facilities. Effective ecotourism, however, requires that protected area personnel be able to work closely and knowledgeably with local people and community leaders as well as with a wide variety of tourism industry representatives including tourism operators, travel agents, tour guides, government tourism agencies and others. Protected area personnel must be able to guide the sometimes conflicting inter- ests of all of the ecotourism participants so that they come together for the benefit of the protected area and its conservation goals. This task is a difficult one but cannot be left to anyone else. In some cases, however, it may be useful for NGOs to assume this role, usually at the request of the protected area administration.

Protected area managers and staff play crucial roles in ecotourism. As the main authorities on their protected area's plants and animals, they provide valuable input to create environmental education programs and impact monitoring systems. On the frontlines of management, protected area personnel are the first to notice natural resource changes such as environmental damage from tourism.

Local communities

People who live in or near protected areas are not a homogeneous group. Indeed, even within one small community there will be a diversity of people with a range of views and experiences. But we can make a few generalizations about local residents and their relationship to ecotourism. First, some rural communities that once featured quiet living are finding themselves in the middle of an international trend. Nature tourists are invading their homelands, but they are generally just passing through the neighborhood, not coming to meet residents.

Residents have mixed reactions to this intrusion. Some want nothing to do with tourists; they want privacy and do not welcome the changes that tourism brings. Others are intrigued by tourism and are taking steps to develop it. Tourism may be particularly alluring if other employment options are limited or if residents feel tourism may help protect their precious resources.

Many communities in developing countries are hosting visitors and creating ecotourism programs. Sometimes their motivation is to protect their surrounding natural resources. For others, they may see ecotourism in a more economic perspective, as a means to gain income. Many communities have organized their own ecotourism programs.

Whatever their initial reaction to tourism, local residents are often unprepared for its demands. Those who do not want tourism have no means to stop it. They often cannot compete with the powerful tourism industry or the fiercely independent travelers who want to discover new areas. Those who are interested in pursuing tourism may not be familiar with its costs and benefits. Many have little experience in tourism business enterprises and are not connected to international tourism markets.

The interests and concerns of local residents regarding tourism development need special attention. Tourism touches all the other groups involved professionally, in a mostly economic sense. For members of communities, it also touches their personal lives by affecting their lifestyles, traditions and cultures, as well as their livelihood and their long standing ways of organizing themselves socially and politically. In addition, most of the other players enter into tourism voluntarily, whereas in many cases communities must deal with tourism impacts whether or not they choose to.

Local residents play an important role in ecotourism for two main reasons. First, it is their homelands and workplaces that are attracting nature travelers. Equity and practicality require that they be active decision-makers in ecotourism planning and management. Second, local residents are key players in conserving natural resources both within and outside of neighboring protected areas. Their relationship to and uses of natural resources will determine the success of conservation strategies for protected areas. In addition, local or traditional knowledge is often a key component of visitors' experience and education.

Tourism industry

The tourism industry is massive. It involves a huge variety of people including: tour operators and travel agents who assemble trips; airline and cruise ship employees; minivan drivers; staff of big hotels and small family lodges; handicraft makers; restaurant owners; tour guides; and all the other people who independently offer goods and services to tourists. The complexity of this sector indicates how challenging it can be for protected area staff and local communities to learn about and form partnerships with the tourism industry.

Consumers are in contact with many members of the tourism industry throughout their journeys. For an international trip, the traveler often first contacts a travel agent, tour operator or airline. The agent will generally contact an outbound tour operator based in the tourist's country of origin, who in turn will contact an inbound tour operator based in the destination country. The inbound tour operator is best placed to make local travel arrangements such as transportation, accommodations, and guide services. Once the traveler is at the destination, many local entrepreneurs will also become part of this scenario.

One element that binds all businesses within the tourism industry is the pursuit of financial profit. There may be additional motivations for some businesses, particularly those involved in ecotourism, but tourism companies exist only when they are profitable.

Members of the tourism industry are valuable to ecotourism for many reasons. First, they understand travel trends. They know how consumers act and what they want. Second, the tourism industry can influence travelers by encouraging good behavior and limiting negative impacts in protected areas. Third, the tourism industry plays a key role in promoting ecotourism. Its members know how to reach travelers through publications, the Internet, the media and other means of promotion, thus providing a link between ecotourism destinations and consumers.

Government officials

Officials from many government departments participate in ecotourism planning, development and management. These departments include tourism, natural resources, wildlife and protected areas, education, community development, finances and transportation. Ecotourism involves officials primarily from the national level, although regional and local levels also contribute to the process.

Government officials have several significant functions in ecotourism. They provide leadership. They coordinate and articulate national goals for ecotourism. As part of their overall tourism plans, they provide vision for this industry. They may even propose a national ecotourism plan; in Australia, the government created a National Ecotourism Strategy and then committed AUS$10 million for its development and implementation.

Government officials at the national level may also establish specific policies for protected areas. For example, government officials decide about visitor use fee systems at protected areas, and their policies outline what systems are established and how revenues will be distributed. They may also delineate private sector practices, e.g., tour operators may be required to use local tour guides in certain areas or developers' property ownership rights may be regulated. Government policies direct ecotourism activities and may easily advance or hinder their growth.

Additionally, government officials are responsible for most basic infrastructure outside protected areas ranging from airline facilities in big cities to secondary roads leading to remote sites. The government generally takes the lead in all major transportation systems and issues. It may also provide other services important to ecotourism such as health clinics in rural areas.

Finally, government officials promote ecotourism. Sometimes the promotion is part of a national tourism campaign. At other times, advertisements for specific nature sites are created or perhaps a flagship species is identified and promoted. National government participation gives prominence to ecotourism destinations.

Nongovernmental organizations. Nongovernmental organizations are valuable players because they provide a forum for discussion and influence regarding ecotourism. They offer a means of communication with great numbers of interested individuals. These organizations can serve as vehicles for bringing together all the elements of ecotourism. NGOs can play many different roles in ecotourism implementation: directly, as program managers or site administrators; and indirectly, as trainers, advisors, business partners with ecotourism companies or communities and, in exceptional circumstances, as providers of ecotourism services.

There are several different types of nongovernmental organizations. Among them are for-profit tourism associations consisting of private tour operators, airlines and hoteliers; ecotourism associations such as those in Belize, Costa Rica, Ecuador, etc., that bring together groups from all the sectors involved; and other trade organizations that handle travel issues. These NGOs often have members who meet regularly and communicate industry concerns through publications such as newsletters. Members are often asked to subscribe to certain principles or "codes of ethics." These associations and organizations are effective at keeping the industry informed about current trends and events.

Another set of nongovernmental organizations involved with ecotourism includes the private, nonprofit groups that focus on conservation and development or may be dedicated specifically to ecotourism. Their focus may be local, national or international. Frequently, these organizations serve as facilitators between protected areas, communities and all the other players in ecotourism, sometimes providing financial and technical assistance or directly managing ecotourism sites. Some of these NGOs have constituencies that enjoy nature and would be interested in ecotourism education and promotion.

Supporting Players

Funders

Many different groups can fund the development of ecotourism through loans or grants: financial institutions, including investment corporations; bilateral and multilateral donor agencies such as the World Bank and the Interamerican Development Bank; private investors; venture capital funds such as the EcoEnterprise Investment Fund; NGOs; and private banks. These contributions are often critical for protected areas that pursue ecotourism. Typically there are studies to carry out, facilities to build, infrastructure to create and people to train. With protected area budgets so limited, outside funding is necessary.

Several international NGOs based in the United States and Europe provide funding and/or technical assistance to ecotourism projects in developing countries. Many of them use funding provided by government agencies such as USAID, GTZ and DFID, the governmental foreign aid departments of the United States, Germany and the United Kingdom, respectively. The Nature Conservancy, through its USAID-funded Parks in Peril program, has helped many local NGOs develop ecotourism projects connected with protected areas. The recently created EcoEnterprise Fund also provides funding on favorable terms for sound ecotourism project proposals

Financial institutions do not generally participate in planning for ecotourism or in decisions about what is appropriate for a particular protected area. In this regard, they may be considered a second-tier player in ecotourism, but they are important nonetheless. For anyone that wants to develop ecotourism, access to funds is often the biggest obstacle confronted.

Academics

Academics at universities is another group that plays a secondary, though valuable, role in the planning and daily functions of ecotourism. It is a group that helps to frame the issues of ecotourism and raise questions to ensure that ecotourism meets its stated goals. Researchers and academics facilitate learning by asking such questions as: Who exactly is benefiting from ecotourism? How do we measure benefits? How does ecotourism contribute to our existing knowledge about conservation? What are the links between ecotourism and tourism? Academics can focus on the big picture and help us understand how ecotourism interacts with other concepts and global trends.

In addition to helping shape the hypotheses, academics conduct research. In coordination with NGOs, governments and local communities, they may:

- develop and execute surveys, e.g., of visitor preferences, willingness to pay, etc.;
- produce data about tourism patterns;
- inventory flora and fauna;
- document tourism impacts and share results to develop a good base of information;
- provide material to guide us in our discussions and conclusions about ecotourism; and
- facilitate the sharing of this information and conceptual thinking through conferences, publications, the Internet, etc.

Travelers

Travelers have a unique position as players in ecotourism. They are the most vital participants in the industry and provide motivation for everyone else's activities, but few participate in formal meetings about ecotourism. Nevertheless, the choices they make when they select a tourism destination, choose a tour operator or travel agent and, ultimately, the kind of tour in which they wish to participate, have a tremendous impact upon the eventual success or failure of ecotourism projects.

Ecotourism, then, is a multifaceted, multi-disciplinary, multi-actor activity requiring communication and collaboration among a diverse range of actors with different needs and interests. Consequently, achieving ecotourism is a challenging process though ultimately enormously rewarding for all involved.

Role of Private and Public Sectors

Despite the many benefits of ecotourism, a number of obstacles need to be overcome. For example, according to a recent survey, respondents indicated that governments can best facilitate ecotourism development by cutting through bureaucratic red tape, providing needed support, and assisting in educational efforts.

Because of the complex development issues—land use and tenure; reliance on imported goods, labor and capital, and lack of sufficient energy and water resources, among others—it is imperative that ecotourism be managed through appropriate policies. These policies need to establish the regulatory and monitoring procedures and assign agency responsibilities in order for ecotourism to flourish. Without these, there is the danger of destroying or damaging the very resources upon which ecotourism relies. In other words, government regulations must find a balance between protection and restriction without hindering individual initiatives.

Both the public and private sectors have important roles to play in making ecotourism work. Generally speaking, governments should provide the necessary support and facilitate private enterprise development, without competing with business ventures. It is helpful to consider the difference between private and public roles.

Role of the Private Sector

The private sector provides the entrepreneurs. They are the key agents in choosing and implementing productive investment opportunities. They are the ones who must take the risks in obtaining the necessary loans to establish ecotourism businesses. They need the necessary financing, managerial skills and knowledge of operations so that they can run their businesses profitably.

Private business decisions must have the public interest in mind, especially since ecotourism means an improvement for society and the environment. However, prospective entrepreneurs may need assistance in learning standard business practices. In the US about 90 percent of new businesses fail within the first few years of operation.

Role of the Public Sector

The challenge for the public sector is to provide a supportive business climate for the private sector, while keeping safeguards to ensure environmental and cultural enhancement. In order to promote ecotourism, government should:

— facilitate efficient private sector activity by minimising market interference and relying on competition as a means of control—even though environmental regulations and monitoring are necessary for ecotourism, government can rely more on incentives and avoid overly restrictive controls that make it difficult for firms to get started or succeed. Further, fiscal and trade policies should promote competition as much as possible;

— ensure a sound macroeconomic environment—governments should support business by helping with grants, loans and incentive programmes, having clear guidelines for business operations, and facilitating compliance with environmental regulations;

— guarantee law and order, and the just settlement of disputes—government should ensure the protection of property rights, while finding ways to resolve differences on desired projects in a fair and equitable manner;

— ensure the provision of appropriate infrastructure—government needs to support ecotourism businesses with roads, water, electricity, waste disposal, etc.;

— ensure the development of human resources—government can do much to encourage certification, education and training programmes through the educational institutions, offices of tourism, and other agencies;

— protect the public interest without obstructing private sector activity with too many regulations

— government should encourage self regulation as much as possible. But where regulations are needed, government can help businesses follow them by providing informational programmes and on-staff specialists for efficient energy use, recycling, design of buildings, etc.;

— promote private sector activity by not competing in the business arena with private enterprise

— Governments have an unfair advantage and a conflict of interest when engaging directly in business ventures. They can start business activities before private firms or individuals have an opportunity to do so. This would create skepticism and dampen private sector development, and thus be damaging for the island economy in the long run;

— acknowledge the role of small business entrepreneurs and facilitate their activities. The role of the private sector is to make the necessary investments that create the ecotourism businesses. Governments must provide a supportive role in establishing the financial and regulatory environment to facilitate the easy entry and smooth operation of ecotourism enterprises. Both need to cooperate in order to achieve the highest standards possible in business management and environmental protection.

Finding a Proper Balance

A number of interrelationships are important when viewing ecotourism within the total picture. Ecotourism is very complex since it involves ecology, culture and business. It is helpful to consider a number of trade-offs between:

— development versus conservation,
— supply versus demand,
— benefits versus costs, and
— people versus environment.

Development versus conservation

Ecotourism policy is not simply passing environmental legislation. Pacific Island residents regard sustainable economic growth and culture as important, if not more important, than protection of the environment. Thus, legislators need to make sure that economic and cultural goals are being met. In fact, policy makers need to seriously consider all three aspects—economic growth, culture, and environment protection—in ecotourism policies.

Supply versus demand

Policy makers need to consider both the supply and the demand aspects of ecotourism. Supply represents ecotourism products like nature parks, conservation or beautification projects, local-style accommodation, ecotours, cultural activities, recreation, resort projects, and small businesses. The provision of eco-products requires decisions and agreements about land and resource use, including zoning, tax incentives, permits, awards programmes, foreign investments, immigration, etc. Governments need to consider establishing a minimum of one nature park for a protected area, and encourage private operators to offer appropriate accommodation and activities for ecotourists.

Demand refers to the ecotourism market—what kind of ecotourist can we attract, how many will visit, where will they stay, how much will they spend, what are they interested in doing, and so on. Decision-makers need to coordinate marketing and promotional efforts so that there is a clear image presented to the consumer about the destination. Governments can also assist with the collection of market information necessary for effective promotions. Furthermore, in order for long term success, there has to be a balance between demand and supply. Too few ecotourists will result in wasted investments, and too many will defeat the purpose of ecotourism. Hence, government needs to balance marketing and development, and not neglect one over the other.

Benefits versus costs

Government needs to ensure that the distribution of benefits and costs are equitable. This requires determination of appropriate visitor fees, adequate returns on investments, sufficient impact fees from developers, optimal returns to local residents, reasonable incentives and taxation. In order for ecotourism to be economically self-sufficient, policy-makers need to make appropriate financial and fiscal policies. In addition, non-economic considerations like social and environmental costs and benefits need to be considered.

People versus environment

The challenge here is to find the balance between traditional ways and modern practices. Island people have not always practiced conservation methods because of a history of abundance of resources in relation to population. However, this can no longer continue as resources become depleted through rapid population increases and destruction by natural disasters.

Ecotourism provides a way to encourage preservation of the environment through reforestation, conservation areas, renewable resources, and recycling. However, these conservation efforts will only be acceptable if they are shown to be consistent with cultural values. Through education programmes, government can do much to educate residents about the need to change any damaging practices like littering, removing artifacts, or over-harvesting plants, animals, or shells. Cultural values that support replenishing limited stocks for future use and generations should be emphasised and encouraged.

In addition, how the land and its resources are being used by tourism operators will determine the attractiveness of the destination to visitors, whether or not they will return, and whether or not they will encourage their friends and relatives to visit. Ecotourists travel with a purpose of experiencing nature and insist that proper environmental practices be upheld. They may even boycott a place if appropriate environmental practices are not followed. Governments can help by providing appropriate environmental guidelines for both visitors and residents.

Government's Leadership Role

Government can do much to improve the environment that encourages business development. In the case of ecotourism, it is imperative that government take a leadership role in creating the economic environment necessary for ecotourism businesses to flourish. It also should encourage and support leaders to emerge from private sector and from within the community to provide the necessary business ventures. By taking the lead in getting everyone to work together, the government can encourage to take the necessary cooperative steps in order for ecotourism to succeed.

Governments' Role in Decision-Making

Private industry, landowners, community leaders, educators, etc. all need to be informed and involved in the decision-making process from the beginning. If anyone is left out, there is a strong possibility that attempts to hinder or stop the project will be made. This will be counterproductive to the good intentions of the policy. Community-based economic development is the key to successful ecotourism.

Coordinated Approach to Ecotourism

Since specific ecotourism policies do not exist at the legislative level, government leaders must realise that this must be done. Furthermore, it is imperative that the process for establishing these policies be initiated now. The longer the delay, the greater will be the difficulty of implementation. A number of general guidelines exist for sustainable tourism development and environmental protection. However, regulation of the environment is a complex area, since there is a lack of clear guidelines and legislative authority.

Implementation of laws usually falls under many departments and agencies, and also various levels of government. For example, some agencies in the US with environmental oversight include the Department of Interior, the Department of Commerce, Parks and Recreation, Coastal Zone Management, the Department of Energy, and the Environmental Protection Agency. Even when islands have their own jurisdiction, there may be poor coordination and confusion about lines of authority.

Furthermore, since each agency is primarily concerned with its own specific jurisdiction, there is always the possibility that ecotourism development may be hindered by overlapping, conflicting, or contradictory measures. Therefore, in order to help ecotourism prosper, it is evident that a proactive approach is needed. Each country should consider establishing an overall administrative body that has the authority to coordinate and implement ecotourism policy. Such a body would be useful in determining the proper balance between the various tradeoffs identified above.

Ecotourism Management Plan (EMP)

This section introduces the six principal management strategies that form the backbone of an Ecotourism Management Plan (EMP). These strategies ensure that tourism activities contribute to the conservation goals of a protected area. However, most of them also have application for ecotourism development in any context, including in those areas that are not formally protected.

As you proceeded through the planning process for an ecotourism site, several ecotourism management strategy ideas may already have occurred to you. These may have been suggested by the Conservation Area Planning (CAP) process or perhaps by activities going on at other sites in the region. The CAP process identifies several vital points: targets or systems (species, natural communities or ecosystems) and the stresses that reduce their ecological viability, sources of those stresses (threats), and the relevant stakeholders. It also eventually identifies the strategies that might be used in order to mitigate or eliminate the existing threats to the biological integrity of the site.

Alternatively, the General Management Planning process may have emphasised the need for establishing discrete visitor use zones or establishing mechanisms for generating income from tourism for site management. To ensure that tourism at a protected area is sustainable, it is necessary to implement a strong and effective management program that involves all stakeholders in dynamic, creative ways. Figure 1 illustrates how diverse ecotourism management strategies contribute to an ecotourism management plan.

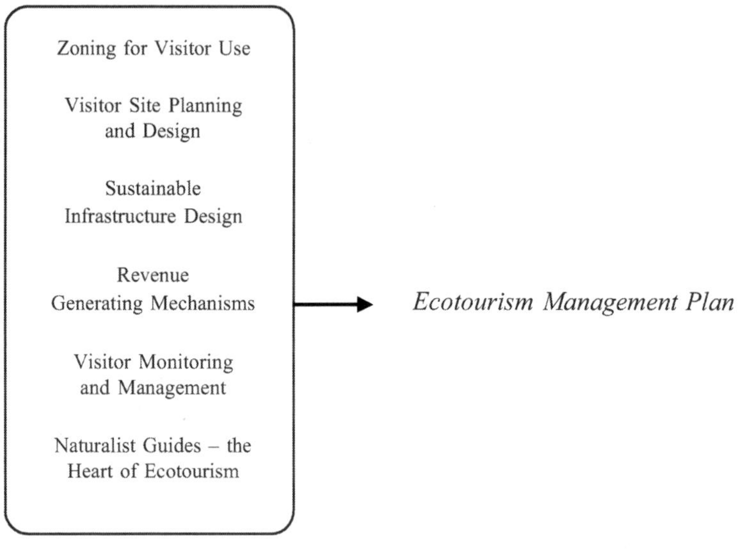

Figure 1. Ecotourism Management Strategies

Zoning for Visitor Use

The appropriate zoning of a protected area is fundamental to all other management strategies. Zoning is a mechanism for assigning overall management objectives and priorities to different areas (zones) within the site or protected area. By assigning objectives and priorities to these zones, planners are also defining what uses will and will not be allowed. These parameters are usually based upon the characteristics of the natural and cultural resource base, the protected area objectives (determined previously) and political considerations. The decision to guide public use using ecotourism principles is a type of political decision that affects zoning. Managers guide their day-to-day decisions about the area's operations based in part upon the zoning structure.

The initial zoning for a protected area is usually determined in the General Management Plan (GMP). However, although ecotourism may be identified in the GMP as the desired public use, current information may be insufficient to define where public use zones should be located. For example, a well-visited waterfall in an area may be an obvious choice for a public use zone in the GMP process, but it may not be until after a Full Site Diagnostic takes place that more worthy attractions outside of preestablished public use zones are identified. Community members and tour operators might, through stakeholder consultations, identify important but previously unexploited attractions, such as a salt lick that attracts parrots on an isolated riverbank.

Consequently, it may be necessary to modify the initial zoning of a protected area following completion of an Ecotourism Management Plan (EMP). Of course, it may be that some potential ecotourism attractions should not be made accessible to visitation because of their vulnerability to erosion or destruction.

Generally speaking, most protected areas provide for two or more types of public use zones. Intensive Use Zones are where most of the high impact, concentrated visitor use takes place, and Extensive Use Zones are where more low impact, generally trail-oriented visitor use occurs. Other zones usually set aside parts of the protected area as "untouchable" zones where very little or no public use occurs, either due to remoteness or resource fragility.

Intensive Use Zones are usually quite small in area, representing less than one percent of a protected area's territory. Extensive Use Zones are generally larger but still represent only a minor part of the site's overall territory. Other zones may permit some ecotourism activities on a highly limited and controlled basis, frequently requiring a permit.

Defining the Zoning Scheme

The first step to defining a zoning scheme is to evaluate the current situation:

— Does the GMP establish a zoning scheme? Is it adequate for the ecotourism objectives that planners have established?

— Can existing or potential negative visitor impacts be eliminated via a good zoning scheme?

— Can existing or potential visitor use conflicts be eliminated via a good zoning scheme?

If the preexisting zoning scheme does not adequately meet the needs for ecotourism development, then changes in the zoning scheme will be needed. The information gathered in the Full Site Diagnostic will enable a refinement of preexisting zoning schemes so that they more accurately reflect the visitor use objectives for the site.

If a protected area's conservation management objectives can continue to be met following the establishment of a proposed visitor site, or if the visitor site's negative impact is outweighed by the benefits it will generate, then it will generally be feasible to overlay preexisting zones with a visitor or public use zone. If conservation management objectives are threatened by establishing a visitor use zone (e.g., if the nesting or feeding area of a rare bird species would be disrupted), then some potentially attractive sites should not be established.

Ecotourism Activities

Ecotourism encompasses a large number of potential activities ranging from ecolodges to trekking. While planning for an ecotourism site, you should decide toward what part of the ecotourism market you wish to orient the site's activities. The wide spectrum of potential ecotourists includes some who will arrive with full understanding of what it means to be ecologically sensitive, while others will need to be educated on site. "High-end" visitors will expect fairly comfortable facilities, while more adventurous or lower spending visitors will seek or settle for more basic facilities.

The type of visitor you wish to have at your site can determine the types of ecotourism activities you plan for as well as the degree to which they are developed. Traditionally, most protected area administrators have opted to manage for a wide variety of visitors, although the facilities they provide generally are geared towards the more basic visitor demands, e.g., campgrounds, trails, small-scale food service. High-end visitors usually find lodging and food service outside the protected area.

As a general rule, high-end visitors spend more money but also require more and better quality facilities that have the potential for causing more environmental impact. The lower-end visitor spends less money but requires only basic services and infrastructure. The more adventurous and lower-end visitor is more likely to utilise sections of the protected area/ecotourism site that are distant and relatively undeveloped.

If ecotourism is to be fully implemented, protected area managers must ensure that tourism activities are low impact and extremely well managed. If these conditions are met, then ecotourism significantly widens the scope and locations for public-use activities. High-end visitor infrastructure may need to be located in a separate zone to avoid possible conflicting uses. Planners and managers must balance the need to generate income with the potential negative impacts and positive economic and educational impacts that can occur with ecotourism.

Remember that a zoning system is not a permanent fixture. Like any plan, it should be modified as conditions change.

Zoning Attributes

When determining zones, one should take into consideration their unique biophysical, social and administrative/management factors. It is a management principle that use in zones not managed for specific attributes will gravitate toward busy, more-developed settings with easier access and a high density of people with corresponding increased evidence of human activity. A well-planned zoning system improves the quality of the visitor experience and provides more options that can enable tour operators to adapt to market changes.

Biophysical Attributes

The natural resources of a zone should be described in terms of their sensitivity and ecological importance. The abundance and density of unique, endangered, endemic or charismatic species that may be important for the zone should be noted.

How natural or intact is the zone, and what evidence of human impact is there? How much scenic beauty is in this zone? What distance from human habitation or difficulty of access is involved? What sorts of human mobility will be allowed?

Social Attributes

Given the biophysical limitations, what type of experience do you wish to offer visitors or other users in the zone?

What user density do you wish to provide? What would be the mix of different types of visitors (e.g., national visitors, international visitors, local people, scientists, etc.)?

What kinds of norms do you expect to govern group movement (e.g., distance, length of stay in visitor sites, waiting time before going to a site, etc.)?

What do you expect to be the group sizes, number of groups per day, types of use and equipment that would be permitted in the zone?

What skill levels would be required before a visitor would be allowed to enter the zone? What are the risks associated with entering the zone?

In zones where local residences are near visitor areas: What are the rules for tourists? Are they allowed to enter the areas where local residents live (i.e., do communities want visitors in their homes and fields)?

Do local residents prefer that their photographs not be taken in these areas (or that a fee be collected for photographing)?

Do local teachers prefer that tourists not visit nearby schools during class time?

In general, what activities are appropriate for the zone? Such zoning can give local residents the ability to control tourist activity so that there is the desired balance of privacy versus interaction.

Administrative Attributes

In order to distinguish between the experiences offered and the permitted uses in the different zones, you must describe the necessary levels of protection and management in each zone and the rules and management actions needed to effectively control the types of activities you wish to have take place there.

What degree of autonomy will visitors have in the zone? Will they need permits? Reservations? Can they leave the trails? Do they need a guide? Can they stay as long as they want? How much patrolling will there need to be in the zone?

What kinds of infrastructure are permitted in the zone? Trash disposal? Signage? Types of trails? Campsites? Campfires?

Zoning Format

After considering the various attributes the zoning scheme should have, the zones should be defined on a map of the area and described. Normally, a zoning scheme includes zones with a range of visitor-use levels. The following format has proven to be a useful one.

— *Name of the zone*: The name should appropriately describe the type of activity that is permitted in the zone, e.g., intensive use, extensive use, primitive use, wilderness, moderate use, etc.

— *General objective*: What are you trying to accomplish with this zone? With regard to ecotourism, what general sort of visitor experience are you trying to provide? How does the zone reflect the site's general management objectives?

— *Zone description*: The description should include a synopsis of the various attributes that will characterise the zone: biophysical, social and administrative.

— *Zone boundaries*: It should describe the location(s) of the particular zone, if possible giving precise boundaries.

— *Management rules, regulations and policies*: Indicate what specific rules, regulations and policies are needed to govern visitor use of the zone, e.g., use of guides, skill levels, permits, camping, use of soap, campfires, group size, etc.

Visitor Site Planning and Design

Site design is a process of intervention involving the sensitive integration of circulation, structures and utilities within natural and cultural landscapes. The process encompasses many steps, from planning to construction.

Most ecotourism sites and protected areas are fairly large covering thousands or tens of thousands of hectares. When planning for ecotourism on a large tract of land or water, visitor use is generally concentrated in a few small sites where most infrastructure is located. Generally referred to as visitor sites, where most visitor use occurs, they are also where some very serious impacts may occur, which is why they must be planned in.

Usually visitor site planning takes place within the context of the preparation of an Ecotourism Management Plan (EMP) and after a zoning scheme for an area has been established. Site plans are prepared as part of the EMP or as a subsequent step when more time and funding are available. Visitor site designation is the result of the EMP process, which analyses natural and cultural resources and attractions of the protected area, makes a determination about the area's ecotourism potential and then selects certain strategic sites for ecotourism concentration based on their:

— inclusion of current and potential ecotourism attractions;

— accessibility;

— potential to concentrate visitor use with a minimum of impact; and/or

— history of previous use. In most cases, it is advisable to use sites that have already received some human intervention in order to avoid impacting intact sites.

The EMP should also make recommendations about the type(s) of infrastructure (e.g., trails, campgrounds, ecolodge, etc.) for the site without being specific about exact locations. The site planning process determines exact locations of infrastructure, taking into account the site's ecological sensitivity and positioning the infrastructure from a visitor management perspective (e.g., location of trails in relation to a campground or attraction). A financial feasibility study will determine whether there is or will be sufficient demand for a business-focused infrastructure (e.g., an ecolodge) and an environmental feasibility study will assess its environmental viability.

The visitor site planning process is best carried out by a team made up of a landscape architect, a biologist or ecologist, and an environmental engineer, who should all have some training in environmental impact evaluation and tourism infrastructure. It is advisable to include a local resident on the team who is familiar with the site and/or environmental conditions in the area.

Funding for this process may be provided by the area's administration or by a potential ecolodge entrepreneur as a part of the cost of developing the business.

Initial Site Planning Considerations

The first step in preparing a visitor site plan is to survey and analyse the proposed location for the recommended infrastructure. It may be necessary to look at a fairly large area and then reduce the effective site's area depending upon results of the analysis.

At this point, the following questions should be asked and answered in at least a provisional manner:

— Is the site appropriate for developing tourism activities according to the General Management Plan (GMP)?

— Can development impacts on the site be minimised?

— What inputs (energy, materials, labour, products) are necessary to support a development option and are required inputs available?

— Can waste outputs (solid waste, sewage effluent, exhaust emissions) be dealt with at acceptable environmental costs?

— What are the potential indicators that should be considered in a future impact monitoring plan for this site?

The next step involves the actual siting of the proposed buildings and infrastructure.

Infrastructure Siting Considerations

When determining exactly where buildings and infrastructure should be located, planners should take into consideration the following:

General Considerations

— Ecosystem maintenance should take precedence over development considerations.

— Plan landscape development according to the surrounding context rather than by overlaying familiar, traditional patterns and solutions.

— Maintain both ecological integrity and economic viability, as both are important factors for a sustainable development process.

— Allow simplicity of functions to prevail while respecting basic human needs of comfort and safety.

— Maximise/minimise exposure to wind through plan orientation and configuration, number and position of wall and roof openings, and relationship to grade and vegetation.

— Recognise that there is no such thing as waste, only resources out of place.

— Assess feasibility of development in long-term social and environmental costs, not just short-term construction costs.

— Plan to implement development in phases to allow for the monitoring of cumulative environmental impacts and the consequent adjustments for the next phase.

Specific Considerations

Capacity: As difficult as it may be to determine, every site has a limit for development and human activity. A detailed site analysis should determine this limit based on the sensitivity of the site's resources, the ability of the land to regenerate and the mitigating factors incorporated into the site's design. The determined limits of acceptable change also depend upon the sensitivity that planners have for the site's environment, and the adaptations that are made to mitigate construction and operational impacts.

Density: Siting of facilities should carefully weigh the relative merits of concentration versus dispersal of visitor use. Natural landscape values may be easier to maintain if facilities are carefully dispersed. Conversely, concentration of structures leaves more undisturbed natural areas.

Slopes: Steep slopes predominate in many park and recreational environments. Siting infrastructure on slopes can cause erosion problems and should be avoided.

Vegetation: It is important to retain as much existing native vegetation as possible to secure the integrity of the site. Natural vegetation is an essential aspect of the visitor experience and should be preserved. Use native species for land generation (not "landscaping"), and avoid the use of exotic plant species. Minimise, or even eliminate, the use of lawns. In areas such as the tropics, most nutrients are held in the forest canopy, not in the soil; loss of trees can therefore causes nutrient loss. Shorelines and beachfronts should not be intensively developed or cleared of vegetation. Vegetation areas should be maintained adjacent to lakes, ponds and streams as filter strips to minimise runoff of sediment and debris.

Buildings and other structures should be sited so as to avoid cutting significant vegetation and to minimise disruption of other natural functions and the natural viewscape. Natural vegetation should be used to diminish the visual impact of facilities and to minimise imposition on the environmental context. In warmer climates, it may be possible to integrate facilities with their environment through minimising solid walls, creating outdoor activity spaces, etc.

Wildlife: Avoid the disruption of movement, nesting patterns, feeding and roosting sites of threatened, endangered or focal wildlife species by sensitive siting of development and by limits set on construction activity and facility operation. Allow opportunities for visitors to be aware of indigenous wildlife (observe but not disturb). Also, be aware that in some ecosystems, particularly on islands, tourism activities can lead to the introduction of invasive species.

Views: Views are critical and reinforce a visitor's experience. Site design should maximise views of natural features and minimise views of visitor and support facilities.

To do so, avoid high structures. Buildings should remain below tree/horizon line and be invisible from the air and on ground arrival as much as possible. Colours used on exteriors should blend, not contrast, with the natural environment.

Natural Hazards: Development should be located with consideration of natural hazards such as precipitous slopes, dangerous animals and plants, and hazardous water areas.

Energy and Utilities: Conventional energy and utility systems are often minimal or nonexistent in potential ecotourism sites. Siting should consider possible connections to off-site utilities or, more likely, spatial needs for onsite utilities.

— Infrastructure should be placed to take advantage of natural ventilation possibilities when consistent with esthetic and other considerations.

— Environmentally appropriate technologies and facilities for the treatment of organic wastes should be considered, such as composting, septic tanks and biogas tanks.

— Provision should be made for ecotourism appropriate facilities that may not have been considered in the original site planning recommendations: facilities for trash storage until removal from the site, solar panels or other appropriate energy source, maintenance buildings and sites for treatment of gray water.

— Water sources should be located where other activities will not impact them and in such a manner that water use will not significantly alter existing watercourses. Waterlines should be located to minimise disruption of earth and adjacent to trails wherever possible.

Visitor Circulation Systems: Infrastructure elements such as lodging and trails should be located to optimise visitor circulation: minimum distances, minimum disturbance to natural features, easily located by visitors, etc. Trails should be designed with environmental and cultural interpretation in mind and with attractions and sensitivity the primary determining factors in placement. Wherever possible, trails should be offered for differing levels of physical ability and should form a closed loop to avoid visitors retracing their steps, thus improving their experience. Trails should be clearly delimited to discourage visitors from leaving them.

Trails and roads should respect travel patterns and habitats of wildlife, including maintaining canopy cover unbroken. They should also conform to existing landforms. Low impact site development techniques such as boardwalks should be used whenever possible instead of paved or unpaved trails; where necessary, they should incorporate erosion controls.

If vehicular access is possible, the extent of roads and other vehicular access routes should be minimised. If a road is needed for supplying the lodge, consider using electric or hybrid vehicles to transport supplies from the main road in order to reduce noise, water and air pollution.

Conflicting Uses: If the site provides for different types of visitor use, for example ecolodge and campground, make sure these uses are sufficiently separated geographically so that they do not conflict.

Safety, visual quality, noise and odour are all factors that need to be considered when siting support services and facilities. These areas need to be separated from public use and circulation areas. Under some circumstances, utilities, energy systems and waste recycling areas can be a positive, educational part of the ecotourism experience.

Siting should be compatible with traditional agricultural, fishing and hunting activities. Some forms of development that supplant traditional land uses may not be responsive to the local economy.

Impact Monitoring: Specific indicators and standards should be established to monitor the impact of the site's use as an ecotourism location.

Sustainable Infrastructure Design

Once a detailed site plan is complete, the task of designing the actual structures and other infrastructure comes next. For most of this infrastructure, e.g., interpretive trails, campgrounds, ecolodges and associated support systems, an architect who is experienced in design of ecotourism projects is needed. This is an extremely important job that should be entrusted only to individuals who truly understand the importance of designing in harmony with natural forces and ecological processes.

Principles of Sustainability

Sustainability does not require a diminished quality of life, but it does require a change in mindset and values toward a less consumptive lifestyle. These changes must embrace global interdependence, environmental stewardship, social responsibility and economic viability.

Sustainable design must use an approach to traditional design that incorporates these changes in mindset. This alternative design approach recognises the impacts of every design choice on the natural and cultural resources of the local, regional and global environments.

Some guiding principles of sustainability include:

— *Recognise interdependence*: The elements of human design interact with and depend on the natural world, with broad and diverse implications at every scale. Expand design considerations to recognise even possible distant effects.

— *Accept responsibility for the consequences of design decisions*: upon the well-being of humans, the viability of natural systems, and their right to coexist.

— *Eliminate the concept of waste*: Evaluate and optimise the full life cycle of products and processes to approach the state of natural systems in which there is no waste.

— *Rely on natural energy flows*: Human designs should, like the living world, derive their creative forces from perpetual solar income. Incorporate this energy efficiently and safely for responsible use.

— *Understand the limitations of design*: No human creation lasts forever and design does not solve all problems. Treat nature as a model and mentor, not an inconvenience to be evaded or controlled.

Sustainable Building Design Philosophy

Sustainable design balances human needs (rather than human wants) with the capacity of the natural and cultural environments. It minimises environmental impacts and the importation of goods and energy, as well as the generation of waste. Any development would ideally be constructed from natural sustainable materials collected on site, generate its own energy from renewable sources such as solar or wind, and manage its own waste.

Sustainable Building Design Objectives

Sustainable building design must seek to:

— Use the building (or non-building) as an educational tool to demonstrate the importance of the environment in sustaining human life.

— Reconnect humans with their environment for the spiritual, emotional and therapeutic benefits that nature provides.

— Promote new human values and lifestyles to achieve a more harmonious relationship with local, regional and global resources and environments.

— Increase public awareness about appropriate technologies and the cradle-to-grave energy and waste implications of various building and consumer materials.

— Nurture living cultures to perpetuate indigenous responsiveness to, and harmony with, local environmental factors.

— Relay cultural and historical understanding of the site with local, regional and global relationships.

Checklist for Sustainable Building Design

Natural Factors

By definition, sustainable design seeks harmony with its environment. To properly balance human needs with environmental opportunities and liabilities requires detailed analysis of the specific site. How facilities relate to their context should be obvious so as to provide environmental education for its users. Although the following information is fairly general, it serves as a checklist of basic considerations to address once specific site data are obtained.

Climate:

— apply natural conditioning techniques to effect appropriate comfort levels for human activities; do not isolate human needs from the environment

— avoid overdependence on mechanical systems to alter the climate

— analyse whether the climate is comfortable, too cold or too hot for the anticipated activities, and then decide on mitigation of the primary climatic components of temperature, sun, wind and moisture that can make the comfort level better.

Temperature:

— temperature is a liability in climates where it is consistently too hot or too cold

— areas that are very dry or at high elevation typically have the characteristic of large temperature swings from daytime heating to night-time cooling, which can be flattened through heavy/massive construction to yield relatively constant indoor temperatures

— when temperature is predominantly too hot for comfort:

    — minimise solid enclosure and thermal mass

    — maximise roof ventilation

    — use elongated or segmented floor plans to minimise internal heat gain and maximise exposure for ventilation

  - separate rooms and functions with covered breezeways to maximise wall shading and induce ventilation
  - isolate heat generating functions such as laundry and kitchens from living areas
  - provide shaded, outdoor living areas such as porches and decks
  - capitalise on night-time temperatures, breezes or ground temperatures
- when climate is predominantly too cold for comfort:
  - consolidate functions into the most compact configuration
  - insulate thoroughly to minimise heat loss
  - minimise air infiltration with barrier sheeting, weatherstripping, sealants and airlock entries
  - minimise entries not oriented towards sun exposure.

Sun:

- sun can be a significant liability in hot climates but is rarely a liability in cold climates
- sun can be an asset in cool and cold climates to provide passive heating
- design must reflect seasonal variations in solar intensity, incidence angle, cloud cover and storm influences
- when solar gain causes conditions too hot for comfort:
  - use overhangs to shade walls and openings
  - use site features and vegetation to provide shading to walls with eastern and western exposure
  - use shading devices such as louvres, covered porches and trellises with natural vines to block sun without blocking out breezes and natural light
  - orient broad building surfaces away from the hot late-day western sun (only northern and southern exposures are easily shaded)
  - use light-coloured wall and roofing materials to reflect solar radiation
  - in tropical climates use shutters and screens, avoiding glass and exposures to direct sunlight

- when solar gain is to be used to offset conditions that are too cold for comfort
    - maximise building exposure and openings facing south (facing north in the southern hemisphere)
    - increase thermal mass and envelope insulation
    - use dark-coloured building exteriors to absorb solar radiation and promote heat gain.

Wind:

- wind is a liability in cold climates because it strips heat away quicker than normal; wind can also be a liability to comfort in hot dry climates when it causes the human body to dehydrate and then overheat
- wind can be an asset in hot, humid climates to provide natural ventilation
    - use natural ventilation wherever feasible; limit air conditioning to areas requiring special humidity or temperature control such as artifact storage and computer rooms;
    - use wind scoops, thermal chimneys or wind turbines to induce ventilation on sites with limited wind.

Moisture:

- moisture can be a liability if it is in the form of humidity, causing such stickiness that one cannot cool evaporatively (cool by perspiring)
    - strategies to reduce the discomfort of high humidity include maximising ventilation, inducing air flow around facilities and venting or moving moisture producing functions such as kitchens and shower rooms to outside areas
- moisture can be an asset by evaporating in hot, dry climates to cool and humidify the air (a natural air conditioning)
    - techniques for evaporative cooling include placing facilities where breezes will pass over water features before reaching the facility and providing fountains, pools and plants.

Other Climatic Considerations

- rainfall can be a liability if any concentrated runoff from developed surfaces is not managed to avoid erosion

— rainfall can be an asset if it is collected off roofs for use as drinking water
— storms/hurricanes/monsoons/typhoons
  — provide or make arrangements for emergency storm shelters
  — avoid development in flood plain or storm surge areas
  — consider wind effects on walls and roofs when designing structures
  — provide storm shutters for openings
  — design facilities to be light enough and of readily available and renewable materials to be safely and economically sacrificed to large storms, or of sufficient mass and detail to prevent loss of life and material.

Topography

— consider building/land interfaces to minimise disturbance to site character, skyline, vegetation, hydrology and soils
— consolidate functions or segment facilities to reduce footprint of individual structures to allow sensitive placement within existing landforms
— use landforms and the sensitive arrangement of buildings to:
  — help diminish the visual impact of facilities
  — enhance visual quality by creating a rhythm of open spaces and framed views
  — orient visitors to building entrances
  — accentuate key landmarks, vistas and facilities.

Water Bodies

— capture views and consider advantages/disadvantages of offshore breezes
— safeguard water from pollutants from the development itself and from the users
— minimise visual impact of development on waterfront zones
  — use building setbacks
  — consider building orientation and materials
  — avoid light pollution

— allow precipitation to naturally recharge ground-water wherever possible.

Pests

— design facilities to minimise intrusion by noxious insects, reptiles and rodents

— ensure that facility operators use natural means for pest control.

Cultural Resources

— archeological and other sites of cultural importance should be respected and not negatively impacted

— understand the local culture and the need to avoid the introduction of socially unacceptable or morally offensive practices

— consult with local indigenous population for design input as well as to foster a sense of ownership and acceptance

— include local construction techniques, materials and cultural considerations in the development of new facilities

— incorporate local expressions of art, handiwork, detailing and, when appropriate, technology into new facility design and interior design.

Sensory Experience

Visual:

— provide visitors with ready access to educational materials and experiences to enhance their understanding and appreciation of the local environment and the threats to it

— incorporate views of natural and cultural resources into routine activities to provide opportunities for contemplation, relaxation and appreciation

— provide visual surprises within the design of facilities to stimulate the educational experience.

Sound:

— locate service and maintenance functions away from public areas

— space lodging units and interpretive stops so that natural and not human sounds dominate

— restrict the use or audio level of unnatural sounds such as radios and televisions.

Smell"

— allow natural fragrances of vegetation to be enjoyed

— direct air exhausted from utility areas away from public areas.

Selection of Building Materials

Selection Priorities

When the source is sustainable:

— natural materials are less energy-intensive and polluting to produce and contribute less to indoor air pollution

— local materials have a reduced level of energy cost and air pollution associated with their transportation and can help sustain the local economy

— durable materials can save on energy costs for maintenance as well as for the production and installation of replacement products.

In selecting building products, it is helpful to prioritise them by origin, avoiding materials from non-renew-able sources. Be careful not to contribute to the local extinction of a particular useful, much-used tree or other building material.

Primary: Materials found in nature such as stone, earth, flora (hemp, jute, reed, cotton), wool and wood.

— Ensure that new lumber is from certified or sustain-ably-managed forests

— use caution that any associated treatments, additives or adhesives do not contain toxins or off-gas volatile organic compounds that contribute to indoor air/atmospheric pollution.

Secondary: Materials made from recycled products such as wood, aluminium, cellulose and plastics.

— Verify that production of material does not involve high levels of energy, pollution or waste

— verify the functional efficiency and environmental safeness of salvaged (reused) materials and products from old buildings

— consider cellulose insulation; it is fireproof and provides a greater R-value per centimetre thickness than fibreglass

— specify aluminium from recycled material; it uses less energy to produce over initial production

— keep alert for new developments; new environmentally sound materials are coming on the market all the time.

Tertiary: Man-made materials (artificial, synthetic, non-renewable) having varying degrees of environmental impact such as plywood, plastics and aluminium.

— Avoid use of materials and products containing or produced with chlorofluorocarbons or hydrochloro-fluorocarbons that deteriorate the environment

— avoid materials that off-gas volatile organic compounds, contributing to indoor air/atmospheric pollution

— minimise use of products made from new aluminium or other materials that are resource disruptive during extraction and a high energy consumer during refinement

— avoid use of concrete and steel.

Energy Management

An ecotourism site has a responsibility to use the most advanced techniques possible to reduce energy consumption, utilise local renewable sources of energy and educate visitors about environmentally responsible energy consumption. Just as a site has primary natural and cultural resources, it has primary renewable energy resources, such as sun, wind and biogas conversion. Solar applications range from hot water preheating to electric power production with photovoltaic cells. Wind-powered generators can provide electricity and pumping applications in some areas. The biogas conversion process reduces gas or electricity costs and eliminates the release of wastewater effluent into water resources. Biogas can be used for water heating, cooking and refrigeration. The availability, potential and feasibility of primary renewable energy resources must be analysed early in the planning process as part of a comprehensive energy plan. The plan must justify energy demand and supply and assess the actual costs and benefits to the local, regional and global environments. Certainly it is best to avoid adding to pressure on the natural environment by avoiding the use of polluting fossil fuels such as diesel and oil.

Water Supply

Sustainable design should respect the water resources with diligence whatever the natural distribution. The challenge of sustainable design applies more to areas where fresh water is not limited than to dry areas where the economics of high-cost water tends to promote wise stewardship. The principles of sustainable design apply without reservation to all types of climates. In a park or ecotourism development, where health considerations are paramount, water issues centre on providing safe drinking, washing, cooking and toilet-flushing water. Pay close attention to issues of water supply and sustainability, impact of use on local communities and sensible economies (e.g., baths vs. showers).

The cornerstone of any domestic water supply programme is conservation. Water conservation includes using water of lower quality such as reclaimed wastewater effluent, gray water or runoff from ground surfaces for toilet flushing or irrigation of the landscape or food crops. These uses do not require the quality of water that is needed for internal consumption, bathing or washing. With the proper type of wastewater treatment and plumbing hardware, sea water can be used as a toilet-flushing medium.

User education and awareness are key to successful water conservation. Visitors should receive interpretation about the source of the water and the types of energy required to process and distribute water at the site. Positive reinforcement should be provided to visitors by informing them of their actual water savings as well as their responsibility in achieving the goal of water conservation. Appropriate signs of high quality material should be put in restrooms to indicate that management places a high priority on water conservation and to confirm goals and expected behaviours of visitors.

Waste Prevention and Management

Preventing pollution in a sensitive setting means thinking through all of the activities and services associated with the facility and planning them in a way that they generate less waste. Waste prevention leads to thinking about materials in terms of reduce, reuse and recycle. The best way to prevent pollution is not to use materials that become waste problems. When such materials must be used, they should be reused onsite. Materials that cannot be directly reused should be recycled.

Solid Waste

Convert biodegradable waste to compost that can be used on site or made available to local food producers. Non-biodegradable wastes should be separated on site and transported to a properly managed site for adequate disposal. This may have the additional benefit of creating additional employment and could provide environmental education and improve local community infrastructure. Use biodegradable detergents, fats, guest soaps and shampoos, etc. Limit the use of disposable plastic containers, utensils and wrappings and advise guests in advance of ecolodge policy.

Sewage

Evaluate the relative impacts and merits of dry toilets, anaerobic bioseptic treatment, aerobic bioseptic treatment and constructed wetlands. Wastewater should be treated to a level acceptable for agriculture and released into an irrigation system for a small garden behind the facility. This accomplishes three goals at once: use of wastewater instead of simply releasing it into the watershed, reduction of local food resources consumption and provision of organic produce.

Visible, Participatory Systems

The "out of sight, out of mind" mentality regarding waste is perpetuated because the systems that deal with waste problems are all behind the scenes and off-limits. An environmentally-sound facility would ensure the visibility of systems to minimise the generation of waste. Such systems require conscious participation by users, visitors and operators, but should not dominate the experience of the visitor at the facility. If each person does his/her share, the facility can be operated in a more environmentally-sound manner. This can also lead to long-term changes in behaviour, benefiting the participant and the Earth.

Training and Maintenance

Waste prevention requires training the operators, including all users of the system, and performing diligent maintenance. Most waste problems are created by lack of attention. Because waste prevention represents a change in the way activities are carried out, it requires an extra effort to ensure that these practices are maintained until they become routine. In situations with high turnover of both employees and visitors, continuous training and education will be essential.

Garbage/Solid Waste Prevention Strategies

Ideally, nothing should be brought into a resource-related development that is neither durable, biodegradable or recyclable.

Pollution Prevention

All refrigeration should be CFC-free. Avoid the use of aerosols. Use more economical and eco-friendly "hand-pumped" materials, which must be biodegradable. Swimming pool back-washes should have chemical removing filters, and consider alternatives to chlorine for pool cleaning. All gasoline and oil tanks should be secured in their own reservoirs to avoid leakage into the surrounding environment, and vehicles' used oils should be collected and shipped out. Minimise light pollution, especially artificial lighting in outdoor areas, to avoid disturbance to wildlife and to keep the stars visible from the lodge. In planning for visitor facilities, a comprehensive design strategy is needed for preventing generation of solid waste.

Revenue-Generating Mechanisms

Natural resource conservation creates a multiplicity of economic benefits for society such as fresh water, clean air, genetic banks, carbon sinks, coastal protection (coral reefs and mangroves), recreation, etc. However, as these benefits have not been allocated a market value, consumers have typically enjoyed them for free. At lower levels of demand in the past, this pattern may have been sustainable. Today, however, the vociferous demand for natural resources and their often unequal distribution means that they and the ecosystem services they provide are increasingly threatened.

Despite their obvious and growing popularity with tourists, recreational opportunities in protected areas are rarely priced adequately. Parks around the world frequently charge a low, or no, price for providing recreational opportunities to the public. Consequently, the demand for access to a protected area often exceeds an area manager's capacity to manage it. The results of over-visitation are sometimes painfully visible at some sites while at others they are more insidious as baseline data on ecosystem health are often nonexistent and it is difficult or impossible to assess how an area has been degraded over time by excessive tourist use.

In developing countries, governments pressured by structural adjustment programs and debt interest payments increasingly limit funding for protected area conservation. In this context, it is essential that protected area systems not subsidize recreation opportunities for foreign nature tourists and access for tour operators.

Income-Generating Mechanisms

A number of relatively simple market-based mechanisms exist to generate tourism revenues for conservation (Table 1). In general, revenue produced by these activities can be described by the following income-collection categories:

Table 1. Types of Fees and Charges in Protected Areas

Fee type	*Description*
Entrance fees	Allows access to points beyond the entry gate.
Admission fees	Collected for use of a facility or special activity, e.g., museum or photography class.
User fees	Fees paid by visitors to use facilities within the protected area, e.g., parking, camping, visitor centres, boat use, shelter use, etc.
Licenses & permits	For private tourism firms to operate on protected area property, e.g., tour operators, guides, transport providers and other users.
Royalties&sales revenue	Monies from sales of souvenirs.
Concession fees	Charges or revenue shares paid by concessionaries that provide services to protected area visitors, e.g., souvenir shops.
Taxes	Such as on hotel rooms, airport use and vehicles.
Leases and rent fees	Charges for renting or leasing park property or equipment.
Voluntary donations	Includes cash, 'in-kind' gifts and labour, often received through 'friends of the park' groups

Entrance Fees

This is a fee charged to visitors in order to enter a protected area or other ecotourism site. It can be collected at the entrance to the site or previously at another administrative centre. It can be charged directly to the visitor or, alternatively, tour operators may purchase tickets in advance so that visitors on organised tours have the fee included in the total cost of their package. Differential fees are common. In developing countries, citizens are typically charged less than foreign visitors are. This is to be encouraged for several reasons:

— Residents of a destination country (i.e., country of site location) are already paying through taxes for protected area conservation;

— Environmental education and recreation objectives of protected areas normally seek to encourage visitation by local people; and

— Foreigners from developed countries are generally willing to pay more for access to protected areas.

A further differential may be made for students who are usually charged an even lower fee. Table 2 shows an example of how privately-managed protected areas in Belize differentiate between local citizens and foreigners.

Table 2. Entrance Fees to Protected Areas Managed by the Belize Audubon Society

		Entrance fees (US$)	
Protected area	*Hectares*	*Belizean*	*Foreigners Citizens*
Guanacaste National Park	20	0.50	2.55
Blue Hole National Park	232	1.00	4.00
Crooked Tree Wildlife Sanctuary	6,475	1.00	4.00
Cockscomb Basin Wildlife Sanctuary	41,278	1.25	5.00
Half Moon Caye National Monument	3,925	1.25	5.00
Tapir Mountain Nature Reserve	2,728	—	—
Shipstern Nature Reserve	8,903	1.00	5.00

Normally, the objective of charging an entrance fee is to increase the funding available for the area's maintenance and development activities. However, the amount of the entrance fee can also be a mechanism for facilitating or limiting visitor access, depending upon the site's particular situation. If a site's administration wishes to limit visitation because of adverse visitor impacts, raising the entrance fee is one way to attempt this objective. However, raising and lowering entrance fees alone does not always have a direct impact on visitor numbers. It may also have unintended consequences, especially if the fee level has not been defined based upon demand. It requires a thorough knowledge of the demand for a site's attractions before the effect of changing the amount of an entrance fee can be reasonably predicted.

Determining Entrance Fee Levels

Ideally, an ecotourism site should have as its objective the generation of enough income to cover its operating expenses plus a surplus to invest in conservation and community development priorities. Achieving this will depend upon a site's importance as a tourism destination and the management and marketing capabilities of the administration and tourism managers.

There are three principal considerations in determining entrance fee levels:

— Willingness to pay for access to a managed area by the visitor. This is determined by surveying visitors to the site. If the entrance fee being charged is not based on willingness to pay, visitors can be asked if it is the right amount and what the maximum is that they would pay. The survey format might provide a range of entrance fee options to choose from.

— Comparison of fees charged at other similar sites in similar circumstances. Remember to allow for differences in natural/cultural attractions, infrastructure development, etc.

— Cover costs associated with provision and maintenance of recreational opportunities. A minimum level of revenue to be generated from entrance fees and other use fees should be enough to properly finance the costs incurred by area management in providing ecotourism opportunities. Very often protected areas contribute to their own problems by undercharging use fees.

Admission Fees

This is a fee collected for the use of a facility or special activity such as a museum or a photography class.

User Fees

This is a fee charged to visitors for the use of a service or a particular opportunity offered by the site that incurs a cost higher than that covered by the entrance fee. (Some sites opt not to charge an entrance fee but instead charge for whatever activities a visitor wishes to participate in.) Examples of this would be charging a fee for parking, visitor centre use or for camping in organised camping or primitive areas.

Licenses or Permits

These are fees charged to tour operators to allow them to manage visitors in protected areas, e.g., charter boat owners in the Galapagos Islands. Typically, they need to be renewed annually and can be used by protected area managers as a means for controlling and limiting access to an area. Additionally, they can be issued to allow the visitor to carry out a specific activity that requires special supervision/management because it is infrequently participated in or because demand for this activity must be rationed, such as backcountry camping or rock climbing. It is common for some activities to be rationed in order to reduce human impact and/or provide for a particular visitor experience such as a high level of solitude. It is a good mechanism for monitoring how many visitors actually carry out certain activities. Fishing is another activity for which a license is frequently required. Guides and tour operators may also need special permission to work within the site, for which a fee is usually charged.

Sales

In many cases, the site's administration or third parties may sell souvenirs, food and other products to visitors within the site. Profit from these sales is another source of income. Especially where sales are concerned, profit must be calculated carefully after deducting all costs, such as of purchasing or manufacturing the product, labour costs, etc. Third parties must also make a profit before the site's administration receives a percentage.

Concessions

This is a mechanism by which third persons provide a service to visitors within an ecotourism site. The most common examples of this are providing lodging and food services to visitors within the site; offering the use of horses, guided tours and boat transportation can also be done via the concession mechanism.

In some ecotourism sites, the administration may choose to carry out all of these services in-house without involving concessions. On the other hand, most ecotourism site managers find that they either do not have the expertise or the investment capital needed to provide these services in a professional manner. This is a decision that each site management will need to make. In any case, a strong and regularly-audited accounting would be necessary to use this option successfully.

Selection of the concessionaries is usually carried out via a bidding process in which the ecotourism site's administration develops the terms of reference and interested parties offer their services, including the amount they are willing to pay for the opportunity to provide the services. In the case of government-managed protected areas, this process can be long and involved. This is an excellent way to involve local people as either owners of the concession, co-concessionaries with a more experienced tour operator or employees of the concessionaire.

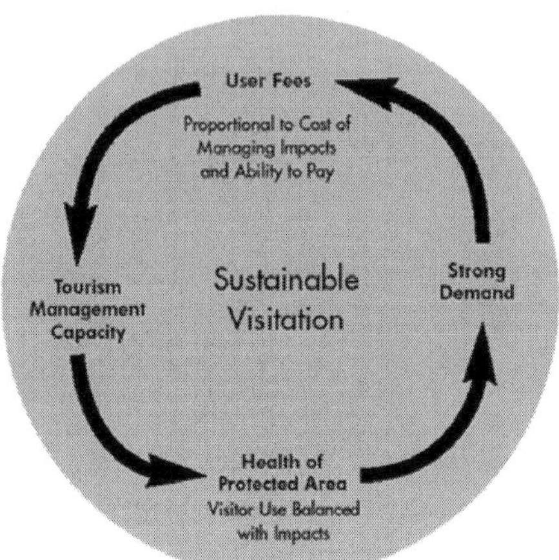

Figure 2. Virtuous Cycle of Tourism User Fees: Positive feedback loop between tourism impacts and conservation finance. A positive feedback loop should exist between fee levels, demand and the health of the protected area ecosystem. Tourism revenues should respond to demand and should possibly used to limit demand in situations where over-visitation is a threat to biodiversity. Income generated from fees should be invested primarily in ensuring tourism's sustainability at the site visited.

A concession may not be a viable alternative for some sites, particularly if there is not much demand for the service. On the other hand, there may be demand but not the entrepreneurs with sufficient capital or interest in taking on the risk of a situation with uncertain results. In any case, a concession should not be undertaken unless a marketing study, business plan and full-scale site plan are prepared.

Concession income can be charged in different ways:

— according to the number of people a concession serves during a given year;
— as a percentage of the gross or net income of the concessionaire;
— as an annual fixed fee; or
— a combination of the above.

In many situations, it is very difficult to calculate profits, income and number of people served by a concession. An annual fee is of course one simple way to charge a concessionaire, but it does not have much flexibility. Remember that a site is supposed to be making money. The concession may annually increase its business while the annual fee stays the same. It is not unusual for concessionaries to make huge profits while site administrations receive very little. It is important to be creative at keeping concession fees appropriate for all but easily calculated. In Costa Rica, the administration of Poas Volcano National Park charges the operators of a coffee shop according to the number of visitors who pay entrance fees.

It should be made clear in the terms of reference that the concessionaire will need to adhere to best practices pertaining to ecotourism infrastructure development and management. For example, standards of cleanliness, maximum numbers of visitors, maximum prices, garbage/trash/human waste disposal should be specified in the concession contract. The ecotourism site's manager, however, is ultimately responsible for ensuring that all standards and contract conditions are monitored periodically and complied with.

Conditions for Collecting Revenues

While there may be many opportunities for generating income in the ecotourism site, producing money requires that you provide the conditions necessary to do so in a safe and professional manner.

Cost/benefit: Just because there is an opportunity to charge visitors for something does not mean that it would be economically justifiable. How much will it cost in order to charge a particular fee? Do you have the personnel available to do this? Will personnel need to place routine but important tasks such as patrolling on the back burner in order to charge an entrance fee? Do you have the infrastructure (e.g., entrance stations) needed to charge the fee? Are there enough visitors to make it worthwhile?

Quality: Visitors will be quick to notice if they are being charged for an inferior product. Before establishing an entrance fee or other fee, be sure that you are offering a product commensurate with the amount of the fee. For example, a high entrance fee should mean that the site offers high value attractions and well-developed and maintained infrastructure as well as sufficient and well-trained personnel. This also applies to concessionaries. Most visitors to the Galapagos National Park in Ecuador are happy to pay the US$100 entrance fee because of the exceptional value of the natural resource and the generally good quality of service they receive. It is important to recognise that income generation should never become an end in itself. You should always keep in mind that your ultimate goal is site conservation. If adding another activity to increase funding for your site is going to interfere with effective long-term site conservation, then you should probably not carry out that activity.

Safety: Because of the location of many ecotourism sites in isolated situations, the safety of the personnel in charge of collecting revenue could be an issue. The safety of the money after being collected could also be a consideration of there is no bank or other secure location for it to be placed until it can be safely deposited in a bank account.

Accounting: The more complex a fee system is, the more important it is to have an appropriate accounting system (and a trained accountant) to adequately administer all of its financial complexities. There are two important reasons for this:

— You need to know exactly how much you are producing from each activity so that you know if it is cost effective. You also need to know how much you are producing in order to develop your next budget (assuming that what you produce can be spent at your site).

— There is a need for transparency and clarity in revenue management. Mismanagement of funds is altogether too common and can be the downfall of a good ecotourism program.

Revenue Distribution

As a general rule, an ecotourism site generates income with a lot more enthusiasm if its personnel know that the income will be spent in large part on the site's management needs. Unfortunately, this is frequently not the case, especially with government-owned protected areas. The majority of income often returns to a general fund where it is used for a wide variety of situations, with very little returning to the site that produced it.

In the United States, both the National Park Service and the National Forest Service have recently begun to allow parks and national forest administrations to retain most of the user and entrance fees that they produce. The Galapagos National Park and Marine Reserve in Ecuador, which produced around US$5 million in 1999, keeps 50% of the fees it generates, while other Galapagos entities, including municipalities, also receive a defined percene.ion.

It may be necessary to lobby the people in charge of financial and budgetary affairs to allow sites to retain a good part of the revenue they produce. In the meantime, being effective, efficient and professional with what you are permitted to do is an important step towards demonstrating that the site's administration should be allowed more freedom to manage its money.

Managing Revenues

If the ecotourism site is allowed to keep all or some of the income that it generates, what should happen to the money once it is collected? An important first step is that it be thoroughly accounted for and deposited in a bank account. If possible, the money should be transferred to a site-focused trust fund. Advantages of using a trust fund include:

— The money will earn interest while it is in the trust fund.

— There is more flexibility to use it than there would be if it were part of a larger institution's administrative structure.

— A select group of individuals may act to oversee the trust fund account and must authorise both investment strategies as well as withdrawals by the site's administration. Frequently, withdrawals must be justified by a work plan presented by the site's administration.

Funding Priorities

In general, income should be spent to ensure that the site meets its conservation objectives. This is a fundamental concept but one that may get lost in the urgency to create a successful ecotourism program. If this cannot be *or* is not done, then the ecotourism program cannot have long term success. There are, however, multiple ways to spend money to meet conservation objectives, and each site must develop its own priorities.

In general, there are three different stakeholder groups that can benefit from the income generated by an ecotourism site: ecosystems, visitors and local people. No matter how the money is spent, or on which group or combination of groups, the bottom line should be conservation. The key conservation benefits of ecotourism can be clustered into five areas:

— A source of financing for biodiversity conservation, especially in legally-protected areas.

— Economic justification for protected areas.

— Economic alternatives for local people to reduce overexploitation on or adjacent to protected areas and other natural areas.

— Constituency building that promotes biodiversity conservation.

— An impetus for private biodiversity conservation efforts.

More specifically, one priority could be ensuring a sufficient flow of funding, i.e., spending money in order to make more money. This could entail building trails, signs, scenic overlooks, etc., to make a site more attractive to visitors. Staff training might also be important. It could also involve doing more to market your site by preparing pamphlets, creating a web site or participating in events where you can publicise a site's attractions.

Perhaps protection of a site's natural resources is a high priority, in which case you might want to hire more personnel, buy more equipment or establish well-defined site boundaries. Another priority is ensuring that visitor impacts are kept to a minimum. Establishing a permanent monitoring program with established procedures and trained personnel is something that all ecotourism sites should have.

If there is an established Ecotourism Program, perhaps the income that is generated should go towards making that program self-sufficient or at least covering its operational budget. Providing local communities with start-up funding to begin an ecotourism enterprise may also be a priority for your site. However a site's priorities are expressed, they should be indicated in the EMP and should be a major factor in determining how ecotourism income will be spent.

Visitor Impact Monitoring and Management

Every time a visitor sets foot in an ecotourism site, he/she causes a negative impact. This is an unavoidable fact. An ecotourism program initiates many public use activities that will have impacts, both positive and negative. An Ecotourism Management Plan seeks to minimise the negative impacts and ensure that they are outweighed by positive ones. The monitoring and managing of visitor impacts are fundamental ecotourism management strategies but ones that are frequently left unattended. If you do not know what effects your ecotourism activities are having on the site's natural environment and the surrounding communities, then you cannot say that you are successful.

Careful monitoring of impacts, both positive and negative, needs to be a primary activity of the site's overall management. Monitoring costs money and requires trained personnel and the assistance of interested stakeholders.

The first methods developed to address tourism impacts evolved from the concept of carrying capacity, which originated in the field of range management. Several definitions of carrying capacity have been offered in the literature depending on how and where the concept was applied. Initially, it was used only to indicate how much tourism activity was too much. Researchers began to realise that looking only at numbers of visitors was not sufficient. They demonstrated that what visitors did, when they did it and a number of other circumstances were frequently more important in determining visitor impacts than simply the number of visitors.

The degree of impact depends upon many variables in addition to the amount of use: the degree of site hardening (making site trails, landings, overlooks resistant to erosion); the motivations and behaviour of visitors; the mode of visitor transport and lodging; the effectiveness of guides; and the season(s) in which most use occurs. Therefore, when managers use the term "carrying capacity" they usually are referring to this more broadly-defined meaning: the amount and type of use that an area can sustain before impacts become unacceptable. The more simple and straightforward concept of carrying capacity limiting numbers of visitors can sometimes be used as a solution for mitigating impacts in restricted, small-scale situations, but not usually on a protected area basis or large ecotourism site situation.

There are two very good methodologies that can be used to monitor visitor impacts: "Measures of Success" and "Limits of Acceptable Change." Limits of Acceptable Change (LAC) has evolved specifically to allow tourism to address the shortcomings of the carrying capacity concept, although it has been applied to more general management situations. Measures of Success can be applied to any management planning situation, not just ecotourism, and relies primarily upon the setting of objectives that can be easily monitored. In order to measure the effectiveness of ecotourism as a conservation strategy, the biodiversity health of the protected area needs to be monitored.

Limits of Acceptable Change Methodology

LAC is a process developed by the United States Forest Service to address visitor impacts, primarily in wilderness situations. It accepts that change is inevitable but sets limits on what degree of change is acceptable. The basic concept involves determining a common vision of what a site's conditions should be, setting indicators and standards related to the amount of change stakeholders deem to be unacceptable in those sites, and then monitoring to continually assess where you are in terms of visitor impacts upon the previously-determined standards. When standards are not met, then management must adapt to mitigate negative impacts. Figure 3 shows a five-step process adapted from Stankey et al.

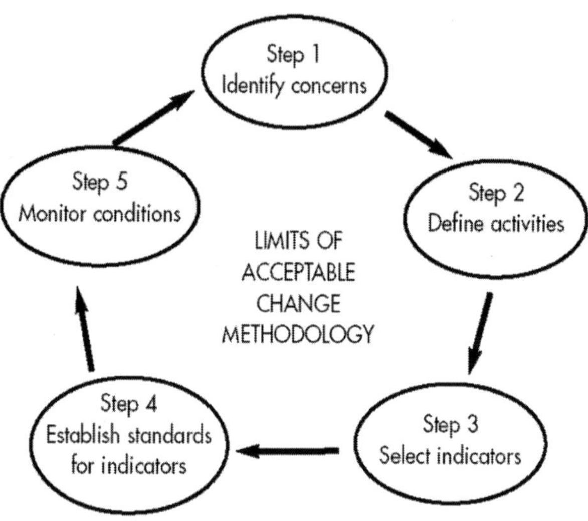

Figure 3. Steps to Implementing Limits of Acceptable Change Methodology

The LAC approach forces managers to come to grips with the details of management in a way that goes far beyond any figure for overall carrying capacity. By setting limits of acceptable change that involve as many stakeholders as possible, managers acquire much more credibility when they request or require management changes that affect other people, such as tour operators, guides and community people.

These are the basic steps in determining the LAC:

— *Identification of Area Issues and Concerns:* Involving all stakeholders, identify the ecotourism site's unique values, attractions, opportunities, threats and problems.

— *Define and Describe the Types of Desirable Activities:* This step should be done in the abstract, not thinking of any specific location. Consider all of the different types of activities that ecotourism might involve. The desirable activities should then be applied to specific sites/zones.

— *Select Indicators:* These indicators should be selected for the management parameters that most concern you at a given site in a given zone. They should be indicators directly related to the activities of visitors that can be controlled.

The following questions should be asked when identifying indicators:

- Does the indicator tell us what we want to know? What question are we trying to answer?
- Does the indicator relate directly to an important resource, social or economic condition?
- Can the indicator be measured easily and relatively inexpensively?
- Can the indicator alert managers to a deteriorating condition before it reaches an unacceptable level?
- Can the indicator be measured without affecting the quality of the visitors' experience?
- Will the indicator provide information that is worth the time and cost needed to obtain it?
- Who will carry out the necessary monitoring?

- *Establish standards for each indicator*: The standards should set some limit of acceptable change. Some impacts are inevitable, but managers must be willing to say how much impact they will tolerate before changing the way they are managing. If trails are eroding faster than it is feasible to maintain them, if viewing areas are getting too big, if some animals are changing their behaviour in an unacceptable way, then management actions must be taken (e.g., group sizes reduced, hardening of some sites, fences put up, patrolling increased).

  Establishing standards requires taking the indicators from the previous step and placing a quantitative value on them: e.g., two landslides per year; 90% of visitors who characterise their visit as "very enjoyable"; two new ecotourism entrepreneurs per year in community X; 25 individual monarch butterflies sighted along trail X between 10 and 11 a.m. on July 20th. Remember that these quantitative values represent limits of some sort that are acceptable; fewer than 90% of visitors who are "very satisfied," or fewer than 25 butterflies sighted along a given trail at a given time, means that managers must determine what is wrong and work to fix it. Establishing indicator standards should involve as many stakeholders as possible so that the standards agreed upon represent everyone's best faith effort and so that they will commit to trying to achieve these limits.

  Some standards and indicators should be chosen from each general type of indicator mentioned above. They should also be chosen for each type of visitor environment, usually by using the zoning system set up in your Ecotourism Management Plan. The types of visitor environment range from intensive use sites where lots of visitors will be found (and there will be high impacts) to primitive and perhaps even wilderness zones, where a high degree of isolation may be desired and managed for (and visitor impact is generally lower).

Another major consideration in choosing standards and indicators is the availability of baseline information. If there is little or no information on which you can base your standards, then you will be making only a very subjective guess about what a realistic standard would be. At first, it may be appropriate to set provisional standards and later adjust them if need be. Bringing in relevant specialists, say a biologist who is familiar with a particularly pertinent species of plant or animal, may help in your decision making.

— *Monitor conditions and implement actions*: If acceptable limits have been exceeded, make management changes that will bring resource, social or economic conditions back within acceptable limits.

The Measures of Success Methodology

The Measures of Success methodology applies the concept of adaptive management and sees monitoring as an essential element of project planning and management. The monitoring program Margoluis and Salafsky describe is integrated into the project cycle and is developed as part of the conceptual model and management plan. Once project goals, objectives and activities are selected, a clear and precise monitoring plan is drawn up. The steps involved in this process are:

— Determining the audiences for monitoring information.

— Determining the information needed based on project objectives (which are prepared so that monitoring can determine whether or not they are being met).

— Designing a monitoring strategy for each information need.

— Developing one or more indicators for each information need.

— Applying and modifying the indicators as needed.

— Determining methods of measuring indicators by using four selection criteria: accuracy/reliability, cost-effectiveness, feasibility and appropriateness.

— Developing an operational plan for applying the methods: listing the tasks, people responsible for carrying out those tasks, monitoring the sites and a timeline for carrying out the plan.

Margoluis and Salafsky provide very detailed information on the types of monitoring designs, the censusing and sampling techniques, the quantitative methods, applying the methods, collecting and handling data, analysing data and communicating results to various types of audiences.

In addition, they have developed another approach for determining project success that can be useful in some ecotourism circumstances. Entitled "Threat Reduction Assessment," this approach identifies and monitors threats in order to assess the degree to which project activities are reducing the threats and achieving success. The process contains the following steps:

— Define the project area spatially and temporally.
— Develop a list of all direct threats to the biodiversity at the project site present at the start date. In the case of an ecotourism project, use the Conservation Area Planning (CAP) results obtained at the beginning of the planning process that identify the major threats to the ecotourism site, and determine strategies for mitigating them.
— Rank each threat based on three criteria: area, intensity and urgency.
— Add up the score for each threat across the three criteria.
— Determine the degree to which each threat has been reduced by management activities.
— Calculate the raw score for each threat.
— Calculate the threat reduction index score.

While natural science methods can be used, less precise social science approaches are often easier to apply, particularly by or about community members/projects. Community members become active participants in future mitigation activities by being involved in this assessment.

Public Participation

While both LAC and Measures of Success require high levels of participation in the planning and operational phases of a monitoring program, Rome recommends the development of a monitoring plan according to a multi-step process that strongly emphasises public participation at practically all levels. According to Rome, the process should be guided by a steering committee composed of protected area/ecotourism site managers, tourism industry representatives and community leaders. It would include the following steps:

— Community meeting to discuss concerns and potential impacts of ecotourism.
— Steering committee meeting to determine indicators and measures and to assign monitoring responsibilities.
— Community meeting to present monitoring program and to discuss limits or ranges of acceptable change.
— Training of monitoring and analysis team.
— Implementation of monitoring.
— Analysis of results, evaluation of management needs and small-scale management adjustments made.
— Community meeting to discuss monitoring results and management recommendations.
— Continued implementation of monitoring and management.

Obtaining the Information

Using management objectives, indicators and standards to assess overall progress requires the ecotourism site's management to have a specific monitoring program that has been incorporated into the site's routine management scheme. Monitoring requires that certain kinds of information be collected on a systematic, routine basis. Baseline information is needed to compare with subsequent data and to assess the direction management is taking.

The collection of baseline data and subsequent data should involve procedures that are relatively simple to implement and do not require large investments of time or cost to the site's administration. To the extent possible, the cost of the monitoring program should be financed from ecotourism revenues.

Most of the data should be collected by the site's staff, but strategic use of third parties such as university biologists, naturalist guides, concessionaries and community members should also be considered. Naturalist guides may also be recruited to carry out certain observations on a routine basis. Cooperative agreements can be signed with local universities that permit scientists to carry out research in return for providing information that will supply baseline data, or to provide data on an ongoing basis that will allow monitoring of a particular management concern. Site staff may need special training to collect certain data. University scientists can train rangers to identify certain insects, bird songs and plants that may be the object of monitoring activity. They can also be trained to take water samples and even do some basic water sample testing.

Some types of data that need to be collected on a daily, systematic basis (which requires a very good recordkeeping system) include: visitor numbers and other visitor characteristics, fee collection amounts, and visitor observations and complaints. In addition, ecotourism management requires frequent evaluation of visitor characteristics and levels of satisfaction with different aspects of the site: facilities, staff, interaction with other visitors, etc. This is usually done using surveys and questionnaires, which can be carried out by site staff or third parties. Ideally, a standard survey addressing the management objectives and indicators of concern should be prepared and presented to a random sample of visitors on a regular basis; alternatively, a select group could be targeted on a more frequent basis, depending upon what is being measured. Visitor comment registers can be placed in strategic places to obtain visitors' opinions. While this is not a scientific method for obtaining visitor input, it can give a sense of what visitors are thinking.

Visitor Management Strategies and Alternatives

If you have determined that you are not reaching management objectives or that you have exceeded a limit of acceptable change, you must adapt your management strategies to this new situation. Table 3 is a framework for considering visitor management strategies.

Table 3. Visitor Management Methods

Indirect Methods	Direct Methods
1. Environmental education/ interpretation	1. Fees and costs
2. Information/diffusion	2. Restrictions
3. Site manipulation	3. Patrolling/human presence
4. Zoning	4. Requirements to participate in certain skilled activities
5. Infrastructure and facility design	5. Permits and licenses
6. Type and degree of maintenance	6. Designated sites (camps,picnics, etc.)
7. Ease or difficulty of access	7. Trained guides
	8. Rules and regulations

Naturalist Guides

Naturalist guides play a central role in the implementation of the ecotourism concept. They are the principal providers of the educational element to the ecotourism activity, and their capacity and commitment ensures that the negative impacts of tourism are minimised. At the same time, guiding is an obvious economic opportunity for people from local communities. These and other important benefits underline the importance of a protected area establishing and implementing a naturalist guide training and licensing program.

The use of tour guides in protected areas is not a new phenomenon. Guides have been a part of nature tourism in many places for many years. They have accompanied tourists on safari in East Africa for several decades. They have travelled with tourists on the boat tours which millions of visitors have enjoyed on the Patagonian lakes of Argentina, particularly in Nahuel Huapi National Park. These tour guides usually were employed by private tour operators and had little or no relationship to the protected area they worked in. Over the years, this situation began to change as protected area managers realised the potential for using guides to increase contact with visitors and for accomplishing other ecotourism objectives as well.

Roles of Naturalist Guides

Naturalist guides truly play a multifaceted role. They have responsibilities to their tour operator employers, to their clients the visitors, and to the protected areas and communities where they work.

Tour operators count on guides to provide experience-enriching interpretation of natural and cultural attractions to add value to the tourists' itinerary. They also require guides to manage logistical aspects of trips in the field, such as coordinating with accommodation, food and transport service providers. Guides are responsible for the tourists' safety and in general represent their tour operator employer in the field.

Tourists look to the naturalist guide for information, interpretation and insight about the places they are visiting; for help preparing for a visit through formal briefings and informal talks; and generally to be a friendly, knowledgeable intermediary with unfamiliar places and people.

Protected area authorities look to the guides as extensions of the park ranger staff, to educate the visitors, to protect the natural and cultural resources of the area visited, to participate in monitoring programs and generally to support the conservation objectives of an area. In addition to these roles, a naturalist guide should seek to inspire visitors to become supporters of conservation.

Nature Interpreters

Environmental interpretation is a subset of communication that focuses on how best to explain environmental/ecological concepts to the general public. One of the central tenets of ecotourism is to educate the visitor. Naturalist guides, who spend a considerable amount of time with visitors, are in a perfect position to educate through skilled interpretation. Many local residents have a detailed knowledge of the plant and animal life as well as of other natural and cultural attractions. They can also relate first-hand experiences with wildlife, medicinal plants and other local phenomena.

Conservationists

As the main contacts that visitors may have with an ecotourism site, guides serve as important role models both to visitors and their own communities. Their attitude and behaviour send an important message to others about the ecotourism concept. Does the guide pick up pieces of trash along the hiking trail? Does the guide actively support and cooperate with site managers by reporting illegal activities? Does the guide adapt ecotourism to his/her own home and community situation? Some tour guides make a point of discussing the importance of conserving the incredible diversity found at a site, what the major threats to it are and what visitors might do to help conserve it.

Park Rangers

Unfortunately, not all visitors to ecotourism sites know how to behave appropriately in sensitive natural and cultural settings. It is the guides' responsibility to ensure that visitors are aware of all applicable rules and regulations as well as other relevant ethical considerations. In a polite but firm manner, they must make sure that visitors comply with whatever restrictions there may be. This is perhaps the most difficult role that guides have because their major responsibility is to help provide visitors with an enjoyable experience. As members of the private sector, it can, in rare situations, create a conflict of interest between the guides' conservation obligations and their obligation to the visitor and, in some cases, their employer. For example, a tour operator might promise clients a close encounter with a whale, but a guide may judge that at a given moment the whales seen in the distance are nursing young and should not be approached. The guide's obligations to an employer and to a park authority might be divergent at this point. Guides need special training in how best to deal with these situations.

Monitors of Tourism Impact

Since guides visit the ecotourism site/protected area on a frequent basis, they are in a unique position to notice certain kinds of impact, such as trail erosion, increasing rareness of a particular bird species, etc. They are also in an excellent position to carry out formal monitoring observations for the site's managers. In many places, guides take the time to carry out observations of the number of nesting birds or of the regeneration of a plant species in a designated quadrant. This can be of valuable assistance to a site's managers when they are short-handed or simply do not have trained personnel to carry out these tasks.

Liaison with Local Communities

When guides are from local communities, they can serve an important role in improving communication between the site's administration and local people. This is particularly important when there may be some misunderstanding between the two different "communities," which there frequently is. Naturalist guides in the Galapagos Islands and other places have established their own organisations to further conservation objectives. In the Galapagos Islands, they have been especially helpful in obtaining local support for the Park Service in the face of illegal fishing activities originating outside the islands.

Conditions for a Successful Naturalist Guide System

In order for a naturalist guide system to work well in an ecotourism site situation, several conditions must be met.

Control and Licensing

The site must have effective control over the use of guides and the conditions under which guides will operate within the site. This implies that managers either own the site or that there is legislation or some other legal mandate for exercising this control. The site's administration reserves the right to suspend or revoke the license if a guide's behaviour is inappropriate. Licenses are usually extended to those individuals who pass a training course or a test. The site's administration reserves the right to set other criteria for attending a training course, such as: being a member of a local community, being of a minimum age, the absence of a police record and having a minimum level of education. It is important to avoid flooding the market with too many licensed guides as this would force down wage levels as many compete for an insufficient number of jobs. However, it is necessary to have a sufficient number of guides to satisfy demand; a rough guide would be to license about 25% more guides than will be working each season.

Mutual Benefits

In spite of the control that the site's administration must exercise over the guides' activities, the relationship between them should be more than one of employer and employee. Both the site administration and the guide have much to offer each other, and they should actively carry out their respective roles in order to benefit from each other's work. Unfortunately, it is not uncommon for one side or the other to lose sight of their mutually supportive roles and for the relationship to become nonproductive. Constant and positive feedback is the best way to avoid this situation. Involving tour operators and guides in the ecotourism program planning process from the beginning is also crucial.

Training

Naturalist guides need training in order to fulfil the many roles they are charged with. The primary themes for a training course curriculum are listed below.

— *Natural history of the site and surrounding areas*: What are the major species, plant and animal communities and ecosystems? How do they interact with one another? What is their conservation status?

— *Cultural attractions*: What are the historical, archaeological and traditional cultural activities that can be found in the site and surrounding areas? What is the relationship between natural and cultural attractions?

— *Site conservation priorities and activities*: Guides should be able to explain to visitors what the site's management is doing to further the conservation of the natural and cultural resources found in the site as well as how the site relates to other protected areas and the surrounding communities.

— *Rules and regulations*: Guides need to be aware of all the rules and regulations governing public use of the site and its facilities. In particular, they need to be aware of what ecotourism is and how it is applied at this site.

— *Group management*: All guides need to learn how to best manage a group of visitors that can have widely varying attention spans and reasons for being there. Maintaining everyone's attention and keeping the group together can sometimes be a major chore. Experienced guides are sometimes the best people to teach this part of the course.

— *Interpretive/communication techniques*: There are very special techniques for communicating ideas to a group of disparate people. Learning the techniques comes easily for some guides; for others, a significant amount of time will need to be spent.

Training should not be a one-time event for guides. Good guides should be continually refreshing and updating their knowledge, and the site's administration should consider carrying out periodic courses for this purpose. Courses should be developed with, and at least partly financed by, the tourism industry. In addition to specialists in each of the themes outlined, tour operators should be instructors in courses, as should older, respected members of the local community.

Guide Availability

Ecotourism encourages the inclusion of local people in as many circumstances as possible. While it may be useful to utilise local people as naturalist guides, managers should realise that residents may not be "natural" naturalist guides. Their interests or educational levels may be obstacles to reaching the level of expertise required of guides at a site. Significant training may be needed before they can function effectively.

Work Availability

Work availability is a very sensitive issue in many situations. Naturalist guides have the potential to earn significantly more money than other members of their community. Managers must be careful not to create high expectations among guide candidates, especially if visitor numbers are not sufficient to guarantee work for everyone. If some candidates for a training course are selected over others who appear to have similar qualifications, conflicts may arise.

Language Skills

Local guides can face a language barrier since most ecotourists are from another country, usually one where a different language is spoken. Local guides can be very ingenious at communicating with visitors whose language they do not speak. However, they cannot express themselves at the level that a high quality naturalist guide would need to communicate effectively, e.g., expressing complex ideas and concepts. When the Galapagos Islands naturalist guide system began, most local guides did not speak any English. Twenty-five years later, almost all of them speak some English or another European language. Some of these guides learned another language on their own by listening and talking to visitors, others took special courses. This ability to communicate in another language has also increased the fee they can charge.

References

Hornback, K.E., and P.F.J. Eagles. (1999). *Guidelines for Public Use Measurement and Reporting at Parks and Protected Areas.* Gland, Switzerland, and Cambridge, U.K.: IUCN.

Isaacs, J.C. (2000). *The limited potential of ecotourism to contribute to wildlife conservation.* The Ecologist. pp. 28(1):61–69.

Patterson, C. (2002). *The business of ecotourism.* Rhinelander, WI.: Explorer's Guide Publishing, pp.40.41, Business Evaluation Criteria.

Weaver, D. B. (1998). *Ecotourism in the less developed world.* C.A.B. Int. Pbl. 288 pp.

Wells, M. P. (1997). *Economic Perspectives on Nature Tourism, Conservation and Development.* Environment Department Paper no. 55, Pollution and Environmental Economics Division.Washington, D.C.: The World Bank.

5

Guidelines for Sustainable Ecotourism

Ecotourism is a concept that evolved over the last 20 years as the conservation community, people living in and around protected areas, and the travel industry witnessed a boom in nature tourism and realized their mutual interests in directing its growth. Ecotourism has brought the promise of achieving conservation goals, improving the well-being of local communities and generating new business — promising a rare win-win-win situation.

Relations among conservationists, communities and tourism practitioners have not always been smooth and collaborative. However, the concept and practice of ecotourism brings these different actors together. Ecotourism has emerged as a platform to establish partnerships and to jointly guide the path of tourists seeking to experience and learn about natural areas and diverse cultures.

Specific circumstances on all sides motivated this new interest in ecotourism. On the conservation side, protected area managers were in the midst of redefining conservation strategies. For practical reasons, they were learning to combine conservation activities with economic development as it became obvious that traditional conservation approaches of strict protectionism were no longer adequate and new ways of accomplishing goals were needed.

For years, conservationists established and managed protected areas[1] through minimal collaboration with the people living in or near these areas. Circumstances in many countries, particularly in developing regions, have changed dramatically in recent years and have affected approaches to conservation.

Over the past two decades, many developing countries have experienced large population increases with declining or stagnant economic conditions. These countries have frequently been pressured into exploiting their natural resource base in an unsustainable fashion in order to meet immediate economic needs and to pay interest on foreign debt. This combination leads more people to compete for fewer natural resources. Outside protected areas, the natural resources that many people have depended upon for sustenance and many businesses have relied upon for profit making have disappeared.

For most countries, protected areas have become the last significant pieces of land that still retain important reserves of plant and animal diversity, water, clean air and other ecological services. Meanwhile, protected areas have become increasingly attractive to farmers, miners, loggers and others trying to make a living. The economic development pressures on these areas have intensified on local, national and global scales. Thus, ecotourism has become very important for potentially reconciling conservation and economic considerations.

Because of this competition for resources, conservationists realized that local people and economic circumstances must be incorporated into conservation strategies. In most cases, local people need financial incentives to use and manage natural resources sustainably. Existing economic and political conditions often limit their options and increase their reliance on natural areas. Conservation work often means creating alternatives to current economic practices so that multiple-use zones around protected areas can be maintained and threats to protected areas minimized.

In looking for alternative economic activities, conservationists have become more creative and are exploring many options. Ecotourism is one such alternative. The rationale behind ecotourism is that local tourism businesses would not destroy natural resources but would instead support their protection. Ecotourism would offer a viable strategy to simultaneously make money and conserve resources. Ecotourism could be considered a "sustainable" activity, one that does not diminish natural resources being used while at the same time generating income.

The explosion in nature tourism has lead to the need to address the impacts of the industry. The growing demand for nature-based tourism sparked interest among protected area managers to place tourism within a conservation context. Travelers have been the driving forces in the evolution of ecotourism. What brought about this nature tourism boom? First, let us examine the status of the tourism industry in general.

According to the World Tourism Organization, world tourism grew by an estimated 7.4 per cent in 2000 — its highest growth rate in nearly a decade and almost double the increase of 1999. Over 698 million people traveled to a foreign country in 2000 spending more than US$476 billion, an increase of 4.5 per cent over the previous year.

The travel and tourism industry supports 200 million jobs worldwide — 1 in every 12.4 jobs. By 2010, this is estimated to grow to 250 million, or 1 in every 11 jobs.

The fastest developing area is East Asia and the Pacific with a growth rate of 14.5%. In the Americas the fastest growth is in Central America (+8.8%).

There is currently no global initiative for the gathering of ecotourism data. However, certain indicators show us how the larger nature tourism market, of which ecotourism is a segment, is growing at a rate faster than that for tourism as a whole, particularly in the tropics.

Ceballos-Lascuráin reports a WTO estimate that nature tourism generates 7% of all international travel expenditure. The World Resources Institute found that while tourism overall has been growing at an annual rate of 4%, nature travel is increasing at an annual rate of between 10% and 30% data which supports this growth rate is found in Lew's survey of tour operators in the Asia-Pacific region who have experienced annual growth rates of 10% to 25% in recent years.

Why are people so attracted to nature destinations? Most likely this trend has followed the global increase in interest in the environment. As people hear about the fragility of the environment, they become more aware of conservation issues around the world. At home, they are willing to pay more for "green" products and services and are taking specific conservation actions such as recycling. For their own pleasure, they want to learn first hand about endangered species and threatened habitats. They want to understand the complex challenges of rainforest conservation and want to experience them first hand.

Travelers are seeking more remote destinations. They want to go off the beaten path, go to the heart of the jungle. Many travelers are becoming activists. As they experience a threatened wilderness area and learn about its plight, they want to help. On the demand side, we have seen a burst of nature tourists seeking new opportunities. International and national travelers are looking for environmental education, are willing to pay entrance fees and are eager to buy local products and services that strengthen the local economy. In sum, they are an ideal audience for addressing the conservation challenges of these areas.

As their interests have changed, consumers have placed new demands on the tourism industry; this, in turn, has encouraged the greening of the tourism industry in addition to encouraging ecotourism. Consumers are requesting new destinations, new ways of doing business and, for some, the opportunity to contribute to natural resource management. Many travel companies are responding to these changing market conditions. Some companies are offering fewer beach vacations and more wilderness treks. New companies devoted solely to nature travel are emerging.

This demand-side trend was destined to coincide with the conservation trend toward effective integration with economic development. When they intersected, people from conservation areas, local communities and the travel sector started talking about ecotourism as a means to meet their common interests. Ecotourism connects travelers seeking to help protected areas with protected areas needing help.

But while the match between conservationists and the tourism industry initially seemed ideal, establishing partnerships has been rocky. Each side continues in the long process of understanding how the other functions and all are learning to incorporate new activities into their work. Ecotravellers — conscious and sensitive nature tourists — constitute a growing segment of the nature tourism market that seeks sensitive interaction with host communities in a way that contributes to sustainable local development. Local communities meanwhile increasingly expect to play a role in the management of tourism.

Working with Ecotourism

A comprehensive view of conservation is implicit in the definition of ecotourism. It incorporates elements of community participation and economic development including the many activities and participants that fulfill this mission.There are many possible ways that ecotourism contributes to conservation. First, ecotourism can generate funds for protected areas. Second, it can create employment for surrounding communities, thus providing economic incentives to support protected areas. Third, it can advance environmental education for visitors. Fourth, it can provide justification for declaring areas as protected or increasing support for these areas. Finally, ecotourism programs aim to limit the negative impacts of nature tourists.

These are the criteria for ecotourism. They provide useful guidelines for judging at what point nature tourism becomes ecotourism. *But this judgement is not simple.* Nor is it an academic or semantic exercise. Only in striving to implement ecotourism and meet all of its criteria in appropriate places will conservation planners and managers meet their long-term goals. We face many challenges in applying these criteria to practical situations in the field.

Actually, implementing ecotourism guidelines is a difficult and complex task. The rewards for a job well done, however, are tremendous. Judgements about ecotourism for a particular site must be done within the context of the area's conservation objectives. As managers and planners investigate actual and potential tourism impacts, both positive and negative, they need to remember the protected area's goals and functions. In some cases, negative impacts from tourism need to be accepted in order to gain conservation benefits. For example, tourism may result in trampled vegetation along trails but also allows for more protected area guards to be hired. Hiring additional protected area guards may be more important to the overall conservation of the protected area than intact vegetation near trails. Whatever the mix of costs and benefits, the key question should be, "Is tourism advancing the long-term conservation agenda of the area?" If so, it is likely ecotourism.

As a final note on the definition of ecotourism, we typically discuss it in the context of protected areas. Protected areas, private reserves and international biosphere reserves are already slated as conservation units and offer the best arenas for pursuing ecotourism.

Although sometimes weak, the legal and management structures of these areas facilitate their ability to capture the benefits and minimize the costs of ecotourism. But ecotourism can take place in areas with less formal conservation status as well. In fact, there may be cases where ecotourism helps establish the protective status of areas currently not formally protected. The rest of this volume and accompanying volumes of this manual are designed to help protected area planners and managers acquire the expertise to navigate successfully among what may appear to be conflicting goals of ecotourism.

Sustainability in the Tourism Sector

Sustainability in the tourism sector was not seriously considered in the past for several reasons which include; firstly the perception that tourism does not involve the consumption of natural resources in the same manner as the agriculture, mining, forestry, fisheries and manufacturing industries do; secondly a mistaken belief that tourism is not a particularly important activity and thirdly; that the sector has not been especially visible nor stakeholders pro-active within the sustainable development debate.

The major sustainability issue related to tourism is to ensure the impact upon the environment is minimised while social and economic goals are achieved as the tourism industry continues to grow. The Precautionary Principle, which states that careful evaluation is necessary whenever practical to avoid serious or irreversible damage to the environment, needs to be considered before any future development. This principle is applied where a range of possible outcomes can not be predicted with confidence, when one or more outcomes could have extreme adverse implication for future generations and where no known substitutes for a resource exists.

Several barriers exist which currently restrict sustainability within the tourism industry. The most important of these is a serious lack of community and industry awareness and education about the issue of sustainability. In addition many tourist sites have a distinct lack of interpretive signage which if present could help create awareness and educate the community. Numerous pockets of research are underway at university and tafe institutions however in many instances this research is conducted in isolation.

Tour operators, licensed through the Department of Conservation and Land Management, often perceive the role of CALM officers as 'environmental policemen' rather than officers willing to provide opportunities to assist them sustain their industry. In addition some operators fail to recognise the need for sustainablity within their own market niche and only see the need for sustainability in the nature based and ecotourism markets. No guidelines are in place to determine the carrying capacity of tourism sites which further creates an air of uncertainty given the seasonal nature of the tourist industry. A monitoring system to assess guidelines and standards of sustainability for current and future tourist sites is also lacking.

Local communities are often not considered in the development of tourism sites and could easily be involved from the outset. Currently several bodies within the tourist industry are approaching the issue of sustainability seriously and seeking ways to support the industry better.

To enhance sustainability within the tourism industry numerous strategies could be adopted and those currently in place refined. The greatest changes could be brought about through education to provide training in sustainable practices for community members, tourism operators and government bodies. Strong partnerships and effective collaboration between interested parties such as government bodies, research institutions, tertiary students, community organisations, and tourist operators if developed further, would enhance sustainability future of the industry. These strategies could comfortably be put into place in the short and medium term provided funding is made available.

Further research into case studies illustrating sustainable practices would provide greater opportunities for newcomers into the industry to follow and demonstrate practical application of the triple bottom line approach. Interpretative signage at tourist sites and public education to encourage members of the community to accept responsibility for the sustainability of sites would further assist. This could be achieved in the medium term perhaps through rewarding responsible behaviour and further development of the 'user contributes' policy introduced by CALM. A framework for licencing all tour operators could be established then strengthened so that realistic sanctions apply if operating conditions are not met.

Different tourism interests may require different types of registration, for example some may be registered as a business, others as individuals or by destination site but each would be required to meet similar conditions, for example to include a prepared management plan, signed memorandums of understanding, and completion of a tour operators course prior to licencing. CALM is currently developing such a course in an online format. Each of these strategies while worthwhile is unlikely to reach its' maximum potential in isolation hence a co ordinated approach is desirable.

Steps have already been taken to provide a comprehensive long term global approach with Australian involvement being achieved through the Cooperative Research Centre for Sustainable Tourism. Together with the World Travel and Tourism Council and other international tourism organisations a world wide programme for sustainable travel and tourism for the 21st century has been established and is known as Green Globe. Green Globe provides a global environmental benchmarking, certification and improvement system for established groups and newcomers to the industry so that triple bottom line requirements, demonstrating environmental, social and economic responsibilities, can be met. Initially companies, communities and protected areas may register as an affiliate to learn more about Green Globe, how to improve sustainability, reduce costs and enhance their green market appeal then to prepare for Benchmarking and Certification.

Prior to Benchmarking, an assessment is completed so that indicators of sustainability, for example energy consumption, waste generation and water consumption, are measured and required guidelines met. Certification is achieved after assessment by independent and accredited audit ('C') companies. Participants are supported in their efforts working towards a best practice level and throughout their involvment, by universities developing further expertise in sustainable tourism strategies and advice from a variety of international and national advisory councils.

Financial commitment in the long term is required from Governments, protective agencies and members of the community to achieve an integrated and systematic approach to sustainable tourism. To obtain some measure of the advantages and disadvantages of the various strategies for change, comprehensive cost-benefit analysis should be conducted. This method of analysis, developed to provide information for use in decision making, can be applied to sustainable tourism where significant savings achieved through Green management and the social impacts of practices which are not sustainable may be measured.

Factors Affecting the Development of Ecotourism

The ecotourism sector of the travel industry is primarily a collection of small- and medium-sized local businesses, communities and non-governmental organisations that develop and implement ecotourism programmes in remote and fragile destinations for both the group tour and independent traveller markets. They are serviced in part by outbound tour operators located primarily in developed countries that specialise in ecotourism. These operators provide the niche marketing and booking services for a significant proportion of organised ecotours worldwide.

For ecotourism to be properly implemented, local and international ecotourism stakeholders are dependent on government to develop policies that will protect and manage natural areas. Government tourism boards and ministries are also crucial players in establishing the reputation and "brand recognition" of the country as an ecotourism destination. Ecotourism stakeholders also depend on the broader tourism industry to transport ecotourists and accommodate them upon arrival in the destination country, or for a part of their stay. After all, many tourists may only spend a portion of their time on an ecotour or in an ecolodge.

Other important stakeholders include local authorities, who often regulate land-use and control key infrastructure, and protected area managers who are responsible for the management of visitors in fragile natural areas. International development agencies also have an important role to play in ecotourism development because they finance projects relating to tourism development, the conservation of biological diversity, and micro-enterprise development—all issues closely related to ecotourism. Finally, successful implementation of ecotourism depends on the development of a stable infrastructure. This includes currency exchange rates, transport systems, peace and security, and good telecommunications systems.

A variety of government ministries are involved with providing a stable environment for business development, and their cooperation is no less important for ecotourism than for other types of international trade. Even though ecotourism businesses are located in remote natural areas, they still require much of the same infrastructure as other businesses to deliver quality experiences for their clients.

Role of Travel Agents

Travel agents play an important role in marketing retail travel industry products but, perhaps surprisingly, have not played a significant role in marketing or sales for the ecotourism industry. In the past, outbound tour operators handled nearly all marketing and retail duties in the ecotourism industry. Now, the internet is increasingly important as a sales medium for ecotourisrn products and is likely to further supplant the need for travel agents in this market niche. Today, travel planning is surging on the internet. In 1999, more than 52 million online travellers used the internet for this purpose—an increase of 54per cent over the previous year.

Role of Outbound Tour Operators

Outbound tour operators are by far the ecotourism industry's dominant marketing and sales organisations. They create the brand name that sells the ecotourism products. They market destinations using four-colour brochures, catalogues with photos of wildlife and ecosystems, World Wide Web pages and, in some cases, through the distribution of film, videotapes and CD-ROM disks. The outbound operator takes responsibility for selecting and packaging the tour product. They must oversee the creation of itineraries to ensure that they will meet the market demand.

Outbound operators handle all sales of the tour product and also handle most air arrangements for their clients through in-house travel agents. They provide tourists with all essential pre-departure information and also are responsible for traveller insurance and liability issues. A common misunderstanding is that outbound tour operators handle all tour details throughout a trip—this is rare. Outbound operators usually contract inbound tour operators to deal with the specifics of a travel programme once the client enters the destination country.

The inbound operator usually represents the outbound operator, making it appear that the "brand-name" company is handling the tours throughout the client's travel experience. Outbound tour operators, however, must also take responsibility for meeting ecotourism objectives as part of their responsibility to oversee their product. This may require extensive work with their inbound operators to ensure that guiding, business, conservation practices and host community relations concur with ecotourism guidelines. It is important to note that while most outbound operators are private businesses, a substantial number are non-profit organisations providing ecotravel as a service for their members.

Role of Inbound Tour Operators

Inbound tour operators usually are located in the major cities of destination countries. They handle multi-day group tours for outbound operators, but they also may offer alternative excursions for walk-in business. Their activities can even extend to providing conference services or customised itineraries directly to individual clients. Inbound operators may also, in some cases, own their own lodges or hotels, which they use for their tour clients. The inbound operator takes the primary responsibility for the client during the trip.

As such, they are the primary entity that is responsible for ensuring that any trip is of a high quality and, particularly, that the educational component meets ecotourism industry standards. To achieve this, inbound companies must have quality interpretive guides. This is the key human resource that establishes an inbound operator as a top competitor in the marketplace, and it is particularly important to their outbound clients.

Ecotourism inbound operators need special support services, such as a network of comfortable but rustic lodges that offer good backcountry experiences and excellent wildlife viewing, while meeting standards of environmental sustainability. They must select ground transportation services that are suitable in size for their small groups, and minimise energy usage and select restaurants that feature local cuisine and are owned by local entrepreneurs. They must also work with local vendors to ensure that tourists have an opportunity to view and ideally purchase genuine local products, such as handicrafts.It is the inbound operator's responsibility to ensure that tourism products generate dollars for local conservation projects. They must work with local communities at each destination site visited to ensure that host communities have proper opportunities to benefit from the tourism programme and that appropriate guest-host interactions exist.

Role of Ecolodges

All ecolodges reflect the creative initiative and entrepreneurialism of business pioneers, rather than large multinational corporations. Found in nature tourism destinations around the world, each ecolodge tends to be individually owned rather than part of a chain. Some lodge owners depend on business from inbound tour operators to provide a steady, predictable client base, while others have established their own market base through direct marketing and public relations strategies. In some cases, skilled entrepreneurs have partnered with indigenous landowners to co-manage the wild land resources that tourists visit and local people depend upon, thereby achieving a positive situation for both the lodge and the local people.

Ecolodges also frequently contribute towards maintaining official protected areas, because they are often quite dependent on proper management of government-controlled reserves. Lodge owners may be involved in long-term agreements with protected areas worldwide, either as concessionaires that pay percentage fees to parks, as "friends" who are financial supporters of park initiatives, or as informal supporters that provide assistance on a project-by-project basis. These ecolodges may actively contribute to trail maintenance, volunteer research programmes, clean-up days, or the monitoring of visitor use, depending on the local situation.

Role of Non-Governmental Organisations

Non-governmental organisations (NGOs) play a prominent role in ecotourism development. They usually are involved for one of two reasons:

i) protection of biodiversity and environment, or

ii) sustainable development for local people.

NGOs are ideal partners for the private sector by developing a myriad of programmes such as research on best practices, guide training, regional planning and stakeholder meetings, community development, protected area management, and targeted conservation initiatives. NGOs also are actively working nationally and internationally to ensure that ecotourism is developing in a manner that is consistent with national and international conservation and sustainable development priorities.

In fact, NGOs worldwide are increasingly developing their own ecotourism programmes because of their strong desire to use ecotourism as a tool for conservation and sustainable development. For example, The Nature Conservancy, the largest private conservation organisation in the world, has developed an ecotourism programme that is assisting with the responsible development of ecotourism products, the planning of ecotourism for protected areas, and the development of user fees to assist with conservation and sustainable development with projects in Ecuador, Belize, Costa Rica, Guatemala, Jamaica, the Dominican Republic, Bolivia and Peru.

Finally, many local NGOs are implementing true grassroots ecotourism initiatives focused on the conservation of local resources that can benefit from ecotourism's economic and educational potential. Turtle-, whale-, penguin- and bird-watching programmes, for example, are an excellent example of how ecotourism can successfully raise awareness and funds for protection, involve local guides and rangers, and lead to long-term sustainable conservation of an endangered species.

Role of Communities

Communities have a vital stake in appropriate ecotourism development in their region, and their participation and involvement are critical to the process. The socioeconomic and cultural impacts of tourism are great, with several well-known negative impacts:

i) loss of local traditions;

ii) commercialisation of local cultural products;

iii) erosion of self-worth;

iv) undermining of family structure;

v) loss of interest (particularly among youth) in land stewardship;

vi) fighting among those that benefit from the tourism cash economy and those that do not;

vii) crime and adoption of illegal underground economies to serve tourists through prostitution, gambling and drugs.

Most researchers agree that some of these social ills can be prevented if the community gives its prior informed consent to any ecotourism projects in its area, participates in tourism development, and remains part of the planning process for tourism entering into the community. If ecotourism is to succeed as a viable form of sustainable development, the private sector, governments and NGOs all must cooperate to include local communities in the development process.

Basic guidelines for community participation offer the approaches required to involve local communities. The need to understand and evaluate a community in advance of proceeding with any partnership or development initiative is paramount, with many assessment approaches available for this purpose. Pre-assessment is important to ensure that community leaders objectively evaluate their needs and goals, and that they are not mistakenly swayed by the offer of funding or business opportunities that might later prove detrimental to the community.

Understanding the decision-making process of a community is highly important. Many local communities do not have a top-down decision-making structure but may instead decide by consensus. This can be very slow and painstaking. It also must be understood that if the community is not involved in initial decisions for the project—such as establishing the objectives for the project or defining its scope—they likely will care less about its success.

Ecotourism development procedures include reviewing with the community the range of possible project' benefits (both economic and social) and its potential negative impacts.

Using trained intermediaries skilled at community facilitation and assessment at this stage is highly recommended. All experts in community development agree that the community must have the information needed to decide if the project's negative impacts may outweigh the benefits, before proceeding with a new project. If the project is approved by the community, community representatives must be clearly integrated into the decision-making process during all phases of the project.

A written agreement between the ecotourism project and the community can help to give both sides the security of having all roles and responsibilities clearly defined from the outset. As part of successful agreements, communities must have the means to invest in projects using the types of resources available to them, such as labour, local renewable resources and land.

Community-Based Ecotourism (CBE) is a growing phenomenon throughout the developing world. The CBE concept implies that the community has substantial control and involvement in the ecotourism project, and that the majority of benefits remain in the community. Three main types of CBE enterprises have been identified. The purest model suggests that the community owns and manages the enterprise. All community members are employed by the project using a rotation system, and profits are allocated to community projects. The second type of CBE enterprise involves family or group initiatives within communities. The third type of CBE is a joint venture between a community or family and an outside business partner.

Role of Indigenous Communities

Ecotourism may often be identified as a means by which communities can raise their standard of living without unsustainable exploitation of natural resources or cultural degradation. Indigenous communities have very special opportunities to develop ecotourism, because they often live in remote natural areas. However, ecotourism is not necessarily the proper course for indigenous communities.

Most indigenous people have been marginalised by their national governments. Frequently they do not have land rights, giving them little clout or control over the development of their homelands. Furthermore, they maybe subject to a number of pressures to change their social, technological and even religious practices to adopt a market- or service-based economy. This can lead to highly unjust situations where tourism is developed without permission on their own lands. The relationship between indigenous communities and tourism has long been tenuous.

Tourism businesses have frequently used local resources with little economic benefit to the community. If local people were involved in tourism, it was as cheap labour or as part of the tourist attraction, mostly in the form of cultural "shows" or displays. This problem continues today, giving indigenous people a natural distrust of tourism development in general, and little reason to believe in the potential of ecotourism. They often perceive it as just one more proposal to eliminate local control over their land and their community's future—and, unfortunately; their concerns may be justified in many instances.

Some indigenous people have been able to take full control of tourism development, while other groups have entered co-management agreements that provide them with a measure of control to protect their land rights, prevent desecration of sacred areas, and reap economic benefits from tourism without undermining their cultural identity. This process cannot proceed successfully unless the indigenous community has legal control over land and full legal rights to protect any businesses that they may establish.

If these vital elements are in place, an indigenous community is in the position to take advantage of ecotourism and use it as a sustainable development tool. Efforts to make ecotourism fully beneficial to local communities are still very new and often experimental. Some communities around the world are keen to get involved, but assistance will be needed to gauge their market potential and real-life business opportunities before development projects are initiated. Certainly communities need their voices heard, and they need to be given the opportunity to develop the skills to fully participate in ecotourism development. Sustained efforts to involve local communities will be required to ensure they are fully part of this international market, when they choose to be.

Role of Women

A wide range of studies on rural development demonstrate that women are less likely to benefit from development initiatives than men, unless special measures are taken to involve them. This problem prevails in ecotourism development projects as well. A study on gender issues and tourism shows that women were willing to increase their workloads tremendously through the sale of handicrafts, in order to gain some new measure of financial independence. Programmes to involve women in ecotourism development are still scarce.

Role of Governments

Governments have an extraordinarily important role to play in the development of ecotourism, yet their role is complex and must be defined by a variety of agency players. Several countries have adopted specific ecotourism strategies. The first national ecotourism plan in the world, prepared by the Australian government in 1994, is the most important example of national ecotourism planning.

The government of Brazil also published Guidelines foran Ecotourism National Policy, which set a benchmark for action in that nation. These plans have a strong focus on the development of appropriate tourism infrastructure and capabilities for the development of tourism in natural areas with a high level of commitment to rural peoples, making them classic examples of ecotourism planning. The commitment to national ecotourism planning development in Australia and Brazil has been very strong because these countries recognised that their tourism economies depend upon the health of their natural ecosystems. Because ecotourism is a growing market, governments around the world are expressing increasing interest in attracting it as part of their tourism development programme.

Role of Development Agencies

Multilateral development agencies, such as World Bank, European Commission and InterAmerican Development Bank (IDB), as well as bilateral agencies including the United Kingdom's Department for International Development (DFID), U.S. Agency for International Development (USAID), German Ministry for Economic Development (GTZ), Norwegian Agency for Development Cooperation (NORAD) and Canadian International Development Agency (CIDA), have become increasingly involved in funding ecotourism projects with loans and grants. Most of these organisations have strict guidelines that target the alleviation of poverty and, in the past, this has kept development agencies and banks out of tourism development work. Since the early 1990s, an ever increasing number of development agency projects have been addressing environmental deterioration and the loss of biological diversity. Small portions of these projects have allocated funding to alternative sustainable development initiatives, such as ecotourisrn. At present, development agencies fund ecotourism development in the following ways:

i) Through programmes that offer conciliatory loan rates to "green" businesses in developing countries.

ii) Through programmes that offer loans to developing nations for the development of tourism as an important source of foreign exchange, with the understanding that tourism must be developed according to strict environmental and social guidelines.

iii) Through loans that seek to use sustainable development of an under-developed region to contribute to the protection of biological diversity of critical ecosystems, such as the Brazilian Amazon.

iv) Through grant programmes that assist the development of micro-enterprises.

v) Through grant programmes that assist the conservation of biological diversity and protected area management.

For the most part, ecotourism is not identified as a funding priority by development agencies, and it is difficult to find it when searching their funding records because it falls under larger categories of development assistance. Because it is such a new category of assistance, few development agencies have specified policies for ecotourism to date. It is clear that new innovative policies for ecotourism are needed to ensure that technical assistance meets the demand for serious long-term projects that are sustainable and fully prepared to compete in the market.

UNEP Principles of Sustainable Tourism

The UNEP (United Nations Environment Programme) principles of sustainable tourism cover the following factors:

1. Integration of tourism into overall policy for sustainable development.
2. Development of sustainable tourism.
3. Management of tourism.
4. Conditions for success.

Integration of Tourism into Overall Policy for Sustainable Development

National Strategies:

Ensure that tourism is balanced with broader economic, social and environmental objectives at national and local level by setting out a national tourism strategy that is based on knowledge of environmental and biodiversity resources, and is integrated with national and regional sustainable development plans.

— Establish a national tourism strategy that is updated periodically and a master plan for tourism development and management.

— Integrate conservation of environmental and biodiversity resources into all such strategies and plans.

— Enhance prospects for economic development and employment while maintaining protection of the environment.

— Provide support through policy development and commitment to promote sustainability in tourism and related activities.

Interagency Coordination and Cooperation:

Improve the management and development of tourism by ensuring coordination and cooperation between the different agencies, authorities and organisations concerned at all levels, and that their jurisdictions and responsibilities are clearly defined and complement each other.

— Strengthen the coordination of tourism policy, planning development and management at both national and local levels.

— Strengthen the role of local authorities in the management and control of tourism, including providing capacity development for this.

— Ensure that all stakeholders, including government agencies and local planning authorities, are involved in the development and implementation of tourism.

— Maintain a balance with other economic activities and natural resource uses in the area, and take into account all environmental costs and benefits.

Integrated Management:

Coordinate the allocation of land uses, and regulate inappropriate activities that damage ecosystems, by strengthening or developing integrated policies and management covering all activities, including Integrated Coastal Zone Management and adoption of an ecosystem approach.

— Maximise economic, social and environmental benefits from tourism and minimise its adverse effects, through effective coordination and management of development

— Adopt integrated management approaches that cover all economic activities in an area, including tourism.

— Use integrated management approaches to carry out restoration programmes effectively in areas that have been damaged or degraded by past activities.

Reconciling Conflicting Resource Uses:

Identify and resolve potential or actual conflicts between tourism and other activities over resource use at an early stage. Involve all relevant stakeholders in the development of sound management plans, and provide the organisation, facilities and enforcement capacity required for effective implementation of those management plans.

— Enable different stakeholders in the tourism industry and local communities, organisations and institutions to work alongside each other

— Focus on ways in which different interests can complement each other within a balanced programme for sustainable development.

Development of Sustainable Tourism

Planning for Development & Land-use at sub-National Level

Conserve the environment, maintain the quality of the visitor experience, and provide benefits for local communities by ensuring that tourism planning is undertaken as part of overall development plans for any area, and that plans for the short-, medium-, and long-term encompass these objectives.

— Incorporate tourism planning with planning for all sectors and development objectives to ensure that the needs of all areas are addressed. (Tourism planning should not be undertaken in isolation.)

— Ensure that plans create and share employment opportunities with local communities.

— Ensure that plans contain a set of development guidelines for the sustainable use of natural resources and land.

— Prevent ad hoc or speculative developments.

— Promote development of a diverse tourism base that is well-integrated with other local economic activities.

— Protect important habitats and conserve biodiversity in accordance with the Convention on Biological Diversity.

Environmental Impact Assessment (EIA):

Anticipate environmental impacts by undertaking comprehensive EIAs for all tourism development programmes taking into account cumulative effects from multiple development activities of all types.

— Examine impacts at the regional national and local levels.

— Adopt or amend legislation to ensure that EIAs and the planning process take account of regional factors, if necessary.

— Ensure that project proposals respond to regional development plans and guidelines for sustainable development.

Planning Measures:

Ensure that tourism development remains within national and local plans for both tourism and for other types of activity by implementing effective carrying capacity programmes, planning controls and management.

— Introduce measures to control and monitor tour operators, tourism facilities, and tourists in any area.

— Apply economic instruments, such as user fees or bonds.

— Zone of land and marine as an appropriate mechanism to influence the siting and type of tourism development by confining development to specified areas where environmental impact would be minimised.

— Adopt planning measures to reduce emissions of CO_2 and other greenhouse gases, reduce pollution and the generation of wastes, and promote sound waste management.

— Introduce new or amended planning or related legislation where necessary.

Legislation & Standards

Legislative Framework:

Support implementation of sustainable tourism through an effective legislative framework that establishes standards for land use in tourism development, tourism facilities, management and investment in tourism.

— Strengthen institutional frameworks for enforcement of legislation to improve their effectiveness where necessary.

— Standardise legislation and simplify regulations and regulatory structures to improve clarity and remove inconsistencies.

— Strengthen regulations for coastal zone management and the creation of protected areas, both marine and land-based, and their enforcement, as appropriate.

— Provide a flexible legal framework for tourism destinations to develop their own set of rules and regulations applicable within their boundaries to suit the specific circumstances of their local economic, social and environmental situations, while maintaining consistency with overall national and regional objectives and minimum standards.

— Promote a better understanding between stakeholders of their differentiated roles and their shared responsibility to make tourism sustainable.

Environmental Standards:

Protect the environment by setting clear ambient environmental quality standards, along with targets for reducing pollution from all sectors, including tourism, to achieve these standards, and by preventing development in areas where it would be inappropriate.

— Minimise pollution at source, for example, by waste minimisation, recycling, and appropriate effluent treatment.

— Take into account the need to reduce emissions of CO_2 and other greenhouse gases resulting from travel and the tourism industry.

Regional Standards:

Ensure that tourism and the environment are mutually supportive at a regional level through cooperation and coordination between States, to establish common approaches to incentives, environmental policies, and integrated tourism development planning.

— Adopt overall regional frameworks within which States may wish to jointly set their own targets, incentive and environmental policies, standards and regulations, to maximise benefits from tourism and avoid environmental deterioration from tourism activities.

— Consider regional collaboration for integrated tourism development planning.

— Develop mechanisms for measuring progress, such as indicators for sustainable tourism.

— Develop regional strategies to address transboundary environmental issues, such as marine pollution from shipping and from land-based sources of pollution.

Management of Tourism

Initiatives by Industry:

Ensure long-term commitments and improvements to develop and promote sustainable tourism, through partnerships and voluntary initiatives by all sectors and stakeholders, including initiatives to give local communities a share in the ownership and benefits of tourism.

— Structure initiatives to give all stakeholders a share in the ownership, to maximise their effectiveness.

— Establish clear responsibilities, boundaries and timetables for the success of any initiative.

— As well as global initiatives, encourage small and medium-sized enterprises to also develop and promote their own initiatives for sustainable tourism at a more local level

— Consider integrating initiatives for small and medium-sized enterprises within overall business support packages, including access to financing, training and marketing, alongside measures to improve sustainability as well as the quality and diversity of their tourism products.

— Market tourism in a manner consistent with sustainable development of tourism.

Monitoring

Ensure consistent monitoring and review of tourism activities to detect problems at an early stage and to enable action to prevent the possibility of more serious damage.

— Establish indicators for measuring the overall progress of tourist areas towards sustainable development.

— Establish institutional and staff capacity for monitoring.

— Monitor the implementation of environmental protection and related measures set out in EIAs, and their effectiveness, taking into account the effectiveness of any ongoing management requirements for the effective operation and maintenance of those measures for protection of areas where tourism activities take place.

Technology

Minimise resource use and the generation of pollution and wastes by using and promoting environmentally-sound technologies (ESTs) for tourism and associated infrastructure.

— Develop and implement international agreements which include provisions to assist in the transfer of Environmentally Sound Technologies (ESTs) for the tourism sector, such as the Clean Development Mechanism of the Kyoto Protocol for energy-related issues.

— Promote introduction and more widespread use of ESTs by tourism enterprises and public authorities dealing with tourism or related infrastructures, as appropriate, including the use of renewable energy and ESTs for sanitation, water supply, and minimisation of the production of wastes generated by tourism facilities and those brought to port by cruise ships.

Compliance Mechanisms

Ensure compliance with development plans, planning conditions, standards and targets for sustainable tourism by providing incentives, monitoring compliance, and enforcement activities where necessary.

— Provide sufficient resources for maintaining compliance, including increasing the number of trained staff able to undertake enforcement activities as part of their duties.

— Monitor environmental conditions and compliance with legislation, regulations, and consent conditions

— Use compliance mechanisms and structured monitoring to help detect problems at an early stage, enabling action to be taken to prevent the possibility of more serious damage.

— Take into account compliance and reporting requirements set out in relevant international agreements.

— Use incentives to encourage good practice, where appropriate.

Conditions for Success

Involvement of Stakeholders

Increase the long-term success of tourism projects by involving all primary stakeholders, including the local community, the tourism industry, and the government, in the development and implementation of tourism plans.

- Involve all primary stakeholders in the development and implementation of tourism plans, in order to enhance their success. (Projects are most successful where all main stakeholders are involved.)
- Encourage development of partnerships with primary stakeholders to give them ownership shares in projects and a shared responsibility for success.

Information Exchange

Raise awareness of sustainable tourism and its implementation by promoting exchange of information between governments and all stakeholders, on best practice for sustainable tourism, and establishment of networks for dialogue on implementation of these Principles ; and promote broad understanding are awareness to strengthen attitudes, values and actions that are compatible with sustainable development.

- Exchange information between governments and all stakeholders, on best practice for sustainable tourism development and management, including information on planning, standards, legislation and enforcement, and of experience gained in implementation of these Principles.
- Use International and regional organisations, including UNEP, can assist with information exchange.
- Encourage development of networks for the exchange of views and information.

Capacity Building

Ensure effective implementation of sustainable tourism, and these Principles, through capacity building programmes to develop and strengthen human resources and institutional capacities in government at national and local levels, and amongst local communities; and to integrate environmental and human ecological considerations at all levels.

— Develop and strengthen their human resources and institutional capacities to facilitate the effective implementation of these Principles.

— Transfer know-how and provide training in areas related to sustainability in tourism, such as planning, legal framework, standards setting, administration and regulatory control, and the application of impact assessment and management techniques and procedures to tourism.

— Facilitate the transfer and assimilation of new environmentally-sound, socially acceptable and appropriate technology and know-how.

— Encourage contributions to capacity-building from the local, national, regional and international levels by countries, international organisations, the private sector and tourism industry, and NGOs.

— Encourage assistance from those involved in tourism in countries which have not yet been able to implement sustainability mechanisms in training at the local and national level in the sustainable development of tourism in co-operation with the Governments concerned.

Charter of Sustainable Tourism

We, the participants at the World Conference on Sustainable Tourism, meeting in Lanzarote, Canary Islands, Spain, on 27-28 April 1995,

Mindful that tourism, as a worldwide phenomenon, touches the highest and deepest aspirations of all people and is also an important element of socioeconomic and political development in many countries.

Recognising that tourism is ambivalent, since it can contribute positively to socio- economic and cultural achievement, while at the same time it can contribute to the degradation of the environment and the loss of local identity, and should therefore be approached with a global methodology.

Mindful that the resources on which tourism is based are fragile and that there is a growing demand for improved environmental quality.

Recognising that tourism affords the opportunity to travel and to know other cultures, and that the development of tourism can help promote closer ties and peace among peoples, creating a conscience that is respectful of the diversity of culture and life styles.

Recalling the Universal Declaration of Human Rights, adopted by the General Assembly of United Nations, and the various United Nations declarations and regional conventions on tourism, the environment, the conservation of cultural heritage and on sustainable development.

Guided by the principles set forth in the Rio Declaration on the Environment and Development and the recommendations arising from Agenda 21.

Recalling previous declarations on tourism, such as the Manila Declaration on World Tourism, the Hague Declaration and the Tourism Bill of Rights and Tourist Code.

Recognising the need to develop a tourism that meets economic expectations and environmental requirements, and respects not only the social and physical structure of destinations, but also the local population.

Considering it a priority to protect and reinforce the human dignity of both local communities and tourists.

Mindful of the need to establish effective alliances among the principal actors in the field of tourism so as to fulfil the hope of a tourism that is more responsible towards our common heritage.

Appeal to the international community and, in particular, *Urge* governments, other public authorities, decisionmakers and professionals in the field of tourism, public and private associations and institutions whose activities are related to tourism, and tourists themselves, to adopt the principles and objectives of the Declaration that follows:

1. Tourism development shall be based on criteria of sustainability, which means that it must be ecologically bearable in the long term, as well as economically viable, and ethically and socially equitable for local communities. Sustainable development is a guided process which envisages global management of resources so as to ensure their viability, thus enabling our natural and cultural capital, including protected areas, to be preserved. As a powerful instrument of development, tourism can and should participate actively in the sustainable development strategy. A requirement of sound management of tourism is that the sustainability of the resources on which it depends must be guaranteed.
2. Tourism should contribute to sustainable development and be integrated with the natural, cultural and human environment; it must respect the fragile balances that characterise many tourist destinations, in particular small islands and environmentally sensitive areas. Tourism should

ensure an acceptable evolution as regards its influence on natural resources, biodiversity and the capacity for assimilation of any impacts and residues produced.

3. Tourism must consider its effects on the cultural heritage and traditional elements, activities and dynamics of each local community. Recognition of these local factors and support for the identity, culture and interests of the local community must at all times play a central role in the formulation of tourism strategies, particularly in developing countries.
4. The active contribution of tourism to sustainable development necessarily presupposes the solidarity, mutual respect and participation of all the actors, both public and private, implicated in the process, and must be based on efficient cooperation mechanisms at all levels: local, national, regional and international.
5. The conservation, protection and appreciation of the worth of the natural and cultural heritage afford a privileged area for cooperation. This approach implies that all those responsible must take upon themselves a true challenge, that of cultural, technological and professional innovation, and must also undertake a major effort to create and implement integrated planning and management instruments.
6. Quality criteria both for the preservation of the tourist destination and for the capacity to satisfy tourists, determined jointly with local communities and informed by the principles of sustainable development, should represent priority objectives in the formulation of tourism strategies and projects.
7. To participate in sustainable development, tourism must be based on the diversity of opportunities offered by the local economy. It should be fully integrated into and contribute positively to local economic development.
8. All options for tourism development must serve effectively to improve the quality of life of all people and must influence the so- cio-cultural enrichment of each destination.
9. Governments and the competent authorities, with the participation of NGOs and local communities, shall undertake actions aimed at integrating the planning of tourism as a contribution to sustainable development.

10. In recognition of economic and social cohesion among the peoples of the world as a fundamental principle of sustainable development, it is urgent that measures be promoted to permit a more equitable distribution of the benefits and burdens of tourism. This implies a change of consumption patterns and the introduction of pricing methods which allow environmental costs to be internalised. Governments and multilateral organisations should prioritise and strengthen direct and indirected aid to tourism projects which contribute to improving the quality of the environment. Within this context, it is necessary to explore thoroughly the application of internationally harmonised economic, legal and fiscal instruments to ensure the sustainable use of resources in tourism.
11. Environmentally and culturally vulnerable spaces, both now and in the future, shall be given special priority in the matter of technical cooperation and financial aid for sustainable tourism development. Similarly, special treatment should be given to zones that have been degraded by obsolete and high impact tourism models.
12. The promotion of alternative forms of tourism that are compatible with the principles of sustainable development, together with the encouragement of diversification represent a guarantee of stability in the medium and the long term. In this respect there is a need, for many small islands and environmentally sensitive areas in particular, to actively pursue and strengthen regional cooperation.
13. Governments, industry, authorities, and tourism-related NGOs should promote and participate in the creation of open networks for research, dissemination of information and transfer of appropriate knowledge on tourism and environmentally sustainable tourism technologies.
14. The establishment of a sustainable tourism policy necessarily requires the support and promotion of environmentally-compatible tourism management systems, feasibility studies for the transformation of the sector, as well as the implementation of demonstration projects and the development of international cooperation programmes.
15. The travel industry, together with bodies and NGOs whose activities are related to tourism, shall draw up specific frameworks for positive and preventive actions to secure sustainable tourism development and establish programmes to support the implementation of such practices.

They shall monitor achievements, report on results and exchange their experiences.

16. Particular attention should be paid to the role and the environmental repercussions of transport in tourism, and to the development of economic instruments designed to reduce the use of non-renewable
17. The adoption and implementation of codes of conduct conducive to sustainability by the principal actors involved in tourism, particularly industry, are fundamental if tourism is to be sustainable. Such codes can be effective instruments for the development of responsible tourism activities.
18. All necessary measures should be implemented in order to inform and promote awareness among all parties involved in the tourism industry, at local, national, regional and international level, with regard to the contents and objectives of the Lanzarote Conference.

Final Resolution

The World Conference on Sustainable Tourism considers it vital to make the following public statements:

1. The Conference recommends State and regional governments to draw up urgently plans of action for sustainable development applied to tourism, in consonance with the principles set out in this Charter.
2. The Conference agrees to refer the Charter for Sustainable Tourism to the Secretary- General of the United Nations, so that it may be taken up by the bodies and agencies of the United Nations system, as well as by international organisations which have cooperation agreements with the United Nations, for submission to the General Assembly.

Berlin Declaration on Biodiversity and Tourism

International Conference of Environment Ministers on Biodiversity and Tourism 6-8 March, 1997, Berlin.

We, Ministers and Heads of Delegation, assembled in Berlin for the International Conference on Biodiversity and Tourism from 6 to 8 March, 1997.

Aware that tourism is an important source of economic wealth and one of the fastest growing sectors in the world economy;

Considering that tourism is a world-wide phenomenon involving a growing number of people undertaking more long-distance journeys;

Recognising that a healthy environment and beautiful landscapes constitute the basis of long-term viable development of all tourism activities;

Observing that tourism increasingly turns to areas where nature is in a relatively undisturbed state so that a substantial number of the world's remaining natural areas are being developed for tourism activities;

Concerned that while tourism may importantly contribute to socio-economic development and cultural exchange, it has, at the same time, the potential for degrading the natural environment, social structures and cultural heritage;

Taking into account that sustainable forms of tourism generate income also for local communities, including indigenous communities, and that their interests and culture require particular attention;

Recognising also that tourism may generate or increase a demand for wild animals, plants or products made thereof for souvenirs, and thus endanger species and affect protection measures;

Further recognising that there is a need to value and protect nature and biological diversity as an essential basis for sustainable development;

Convinced that nature has an intrinsic value which calls for the conservation of species, genetic and ecosystem diversity to ensure the maintenance of essential life support systems;

Furthermore convinced that sustainable forms of tourism have the potential to contribute to the conservation of biological diversity outside and inside protected areas;

Bearing in mind that vulnerable areas, including small islands, coasts, mountains, wetlands, grasslands and other terrestrial and marine ecosystems and habitats of outstanding beauty and rich biological diversity, deserve special measures of protection;

Convinced that achieving sustainable forms of tourism is the responsibility of all stakeholders involved, including government at all levels, international organisations, the private sector, environmental groups and citizens both in tourism destination countries and countries of origin;

Determined to work together with all who are involved in the elaboration of international guidelines or rules that harmonise the interests of nature conservation and tourism, that lead towards sustainable development of tourism, and, thus, contribute to the implementation of the Convention on Biological Diversity and the objectives of Agenda 21;

Agree on the following principles:

1. General

1. Tourism activities should be environmentally, economically, socially and culturally sustainable. Development and management of tourism activities should be guided by the objectives, principles and commitments laid down in the Convention on Biological Diversity.
2. Tourism activities which directly or indirectly contribute to the conservation of nature and biological diversity and which benefit local communities should be promoted by all stakeholders.
3. To conserve nature and biological diversity as a major resource of tourism activities, all necessary measures should be taken to ensure that the integrity of ecosystems and habitats is always respected. Additional burdens from tourism development should be avoided in areas where nature is already under pressure from tourism activities. Preference should be given to the modernisation and renovation of existing tourism facilities.
4. Measures inspired by the principle of precautionary action should be taken to prevent and minimise damage caused by tourism to biological diversity. Such measures should include monitoring of existing activities and assessment of environmental impacts of proposed new activities, including the monitoring of the negative effects of wildlife viewing.
5. Tourism activities which use environmentally sound technologies for saving water and energy, prevent pollution, treat waste water, avoid the production of solid waste and encourage recycling should be promoted to the fullest extent.

   Similarly, tourism activities which encourage the use of public and non-motorised transport should be supported wherever possible.
6. All stakeholders including governments, international organisations, the private sector and environmental groups should recognise their common responsibilities to achieve sustainable forms of tourism.

Policies and, where appropriate, legislation, environmental economic instruments and incentives should be developed to ensure that tourism activities meet the needs of nature and biological diversity conservation, including mobilising funding from tourism

The private sector should be encouraged to develop and apply guidelines and codes of conduct for sustainable tourism.

All stakeholders should cooperate locally, nationally and internationally to achieve a common understanding on the requirements of sustainable tourism. Particular attention should be given to trans boundary areas and areas of international importance.

7. Concepts and criteria of sustainable tourism should be developed and incorporated in education and training programs for tourism professionals. The general public should be informed and educated about the benefits of protecting nature and conserving biodiversity through sustainable forms of tourism. Results of research and concepts of sustainable tourism should be increasingly disseminated and implemented.

2. Specific

1. Inventories of tourism activities and attractions should be developed, taking into account the impacts on ecosystems and biological diversity. Coordinated efforts of governments, the private sector and all other stakeholders should be undertaken to agree on criteria to measure and assess the impacts of tourism on nature and biological diversity. In this regard, technical and scientific cooperation should be established through the clearing house mechanism of the Convention on Biodiversity.
2. Tourism activities, including tourism planning, measures to provide tourism infrastructure, and tourism operations, which are likely to have significant impacts on nature and biological diversity should be subject to prior environmental impact assessments.
3. Tourism activities should be planned at the appropriate levels with a view to integrate socio-economic, cultural and environmental considerations at all levels. Development, environment, and tourism planning should be integrated processes. All efforts should be made to ensure that integrated tourism plans are implemented and enforced.

4. Tourism should be based on environmentally friendly concepts and modes of transport. Negative impacts of transport on the environment should be reduced, paying particular attention to environmental impacts of road and air traffic, specifically in ecologically sensitive areas.
5. Sports and outdoor activities, including recreational hunting and fishing, particularly in ecologically sensitive areas, should be managed in a way that they fulfill the requirements of nature and biological diversity conservation and comply with the existing regulations on conservation and sustainable use of species.
6. Special care should be taken that living animals and plants, and products made thereof for souvenirs, are offered for sale only on the basis of a sustainable and environmentally sound use of the natural resources and in conformity with national legislation and international agreements.
7. Whenever possible and appropriate, economic instruments and incentives including awarding of prizes, certificates and eco-labels for sustainable tourism should be used to encourage the private sector to meet its responsibilities for achieving sustainable tourism. The abolition of economic incentives encouraging environmentally unfriendly activities should be strived for.
8. Tourism should be developed in a way so that it benefits the local communities, strengthens the local economy, employs local workforce and wherever ecologically sustainable, uses local materials, local agricultural products and traditional skills. Mechanisms, including policies and legislation should be introduced to ensure the flow of benefits to local communities.

   Tourism activities should respect the ecological characteristics and capacity of the local environment in which they take place. All efforts should be made to respect traditional lifestyles and cultures.
9. Tourism should be restricted, and where necessary prevented, in ecologically and culturally sensitive areas. All forms of mass tourism should be avoided in those areas. Where existing tourism activities exceed the carrying capacity, all efforts should be made to reduce negative impacts from tourism activities and to take measures to restore the degraded environment.

10. Tourism in protected areas should be managed in order to ensure that the objectives of the protected area regimes are achieved. Wherever tourism activities may contribute to the achievement of conservation objectives in protected areas, such activities should be encouraged and promoted, also as cases to test in a controlled manner the impact of tourism on biodiversity. In highly vulnerable areas, nature reserves and all other protected areas requiring strict protection, tourism activities should be limited to a bearable minimum.
11. In coastal areas all necessary measures should be taken to ensure sustainable forms of tourism, taking into account the principles of integrated coastal area management. Particular attention should be paid to the conservation of vulnerable zones, such as small islands, coral reefs, coastal waters, mangroves, coastal wetlands, beaches and dunes.
12. Tourism in mountain areas should also be managed in environmentally appropriate ways. Tourism in sensitive mountain regions should be regulated so that the biological diversity of these areas can be preserved.
13. In all areas where nature is particularly diverse, vulnerable and attractive, all efforts should be made to meet the requirements of nature protection and biological diversity conservation. Particular attention should be paid to the conservation needs in forest areas, grasslands, fresh water eco-systems, areas of spectacular beauty, arctic and antarctic eco-systems.
14. The Ministers gathered in Berlin on 7 and 8 March, 1997, for the International Conference on Biodiversity and Tourism
    - Recommend that the Conference of the Parties to the Convention on Biological Diversity draw up in consultation with stakeholders guidelines or rules for sustainable tourism development on a global level on the basis of the "Berlin Declaration" in order to contribute to the implementation of the Convention's objectives,
    - Agree to submit the "Berlin Declaration" to all Parties and Signatory States with the objective of bringing about a discussion at the 4th Conference of the Parties in Bratislava,
    - Call upon the Special Session of the General Assembly of the United Nations to support this initiative under the Biodiversity Convention and recommend to the UN General Assembly Special Session to include the subject of sustainable tourism in the future

work program of the Commission on Sustainable Development in order to draw increased attention to the objectives of Agenda 21,

— Call on the bilateral and multilateral funding organisations to take into account the principles and guidelines of the "Berlin Declaration" when supporting projects relating to tourism.

Agreed at Berlin, on the 8th of March, 1997.

Responsible Tourism in Destinations: The Cape Town Declaration, 2002

Shaping Sustainable Spaces into Better Places

We, representatives of inbound and outbound tour operators, emerging entrepreneurs in the tourism industry, national parks, provincial conservation authorities, all spheres of government, tourism professionals, tourism authorities, NGOs and hotel groups and other tourism stakeholders, from 20 countries in Africa, North and South America, Europe and Asia; having come together in Cape Town to consider the issue of Responsible Tourism in Destinations have agreed this declaration.

Mindful of the debates at the United Nations Commission on Sustainable Development in 1999, which asserted the importance of the economic, social and environment aspects of sustainable development and of the interests of indigenous peoples and local communities in particular.

Recognising the global challenge of reducing social and economic inequalities and reducing poverty, and the importance of New Partnership for Africa's Development (NEPAD) in the process.

Recognising the importance of the World Tourism Organisation's Global Code of Ethics, which aims to promote responsible, sustainable and universally accessible tourism and sharing its commitment to equitable, responsible and sustainable world tourism and its STEP initiative with UNCTAD, which seeks to harness sustainable tourism to help eliminate poverty.

Conscious that we are now ten years on from the Rio Earth Summit on Environment and Development, and that the World Summit on Sustainable Development taking place in Johannesburg will put renewed emphasis on sustainability, economic development, and in particular on poverty reduction.

Aware of the World Tourism Organisation, World Travel and Tourism Council and the Earth Council's updated Agenda 21 for the Travel and Tourism Industry and the success achieved by a number of businesses, local communities and national and local governments in moving towards sustainability in tourism.

Aware of the work of the UNEP, and the Tourism Industry Report 2002, and work of UNESCO, and other UN agencies, promoting sustainable tourism in partnership with the private sector, NGOs, civil society organisations and government.

Aware of the guidelines for sustainable tourism in vulnerable ecosystems being developed in the framework of the Convention on Biological Diversity.

Conscious of developments in other industries and sectors, and in particular of the growing international demand for ethical business, and the adoption of clear Corporate Social Responsibility (CSR) policies by companies, and the transparent reporting of achievements in meeting CSR objectives in company annual reports.

Recognising that there has been considerable progress in addressing the environmental impacts of tourism, although there is a long way to go to achieve sustainability; and that more limited progress has been made in harnessing tourism for local economic development, for the benefit of communities and indigenous peoples, and in managing the social impacts of tourism.

Endorsing the Global Code of Ethics and the importance of making all forms of tourism sustainable through all stakeholders taking responsibility for creating better forms of tourism and realising these aspirations.

Relishing the diversity of our world's cultures, habitats and species and the wealth of our cultural and natural heritage, as the very basis of tourism, we accept that responsible and sustainable tourism will be achieved in different ways in different places.

Accepting that, in the words of the Global Code of Ethics, an attitude of tolerance and respect for the diversity of religious, philosophical and moral beliefs, are both the foundation and the consequence of responsible tourism.

Recognising that dialogue, partnerships and multi-stakeholder processes—involving government, business and local communities—to make better places for hosts and guests can only be realised at the local level, and that all stakeholders have different, albeit interdependent, responsibilities; tourism can only be managed for sustainability at the destination level.

Conscious of the importance of good governance and political stability in providing the context for responsible tourism in destinations, and recognising that the devolution of decision making power to democratic local government is necessary to build stable partnerships at a local level, and to the empowerment of local communities.

Aware that the management of tourism requires the participation of a broad range of government agencies and particularly at the local destination level.

Recognising that in order to protect the cultural, social and environmental integrity of destinations limits to tourism development are sometimes necessary.

Having, during the Cape Town Conference, examined the South African Guidelines for Responsible Tourism, tested them in a series of field visits, and explored how tourism can be made to work better for local communities, tourists and businesses alike, we recognise their value in helping to shape sustainable tourism in South Africa.

Recognising that one of the strengths of the South African Guidelines for Responsible Tourism is that they were developed through a national consultative process, and that they reflect the priorities and aspirations of the South African people.

Recognising that Responsible Tourism takes many forms, that different destinations and stakeholders will have different priorities, and that local policies and guidelines will need to be developed through multi-stakeholder processes to develop responsible tourism in destinations.

Having the following characteristics, Responsible Tourism:

— minimises negative economic, environmental, and social impacts;
— generates greater economic benefits for local people and enhances the well-being of host communities, improves working conditions and access to the industry;
— involves local people in decisions that affect their lives and life chances;
— makes positive contributions to the conservation of natural and cultural heritage, to the maintenance of the world's diversity;

— provides more enjoyable experiences for tourists through more meaningful connections with local people, and a greater understanding of local cultural, social and environmental issues;

— provides access for physically challenged people; and

— is culturally sensitive, engenders respect between tourists and hosts, and builds local pride and confidence.

We call upon countries, multilateral agencies, destinations and enterprises to develop similar practical guidelines and to encourage planning authorities, tourism businesses, tourists and local communities – to take responsibility for achieving sustainable tourism, and to create better places for people to live in and for people to visit.

We urge multilateral agencies responsible for development strategies to include sustainable responsible tourism in their outcomes.

Determined to make tourism more sustainable, and accepting that it is the responsibility of all stakeholders in tourism to achieve more sustainable forms of tourism, we commit ourselves to pursue the principles of Responsible Tourism.

Convinced that it is primarily in the destinations, the places that tourists visit, where tourism enterprises conduct their business and where local communities and tourists and the tourism industry interact, that the economic, social and environmental impacts of tourism need to be managed responsibly, to maximise positive impacts and minimise negative ones.

We undertake to work in concrete ways in destinations to achieve better forms of tourism and to work with other stakeholders in destinations. We commit to build the capacity of all stakeholders in order to ensure that they can secure an effective voice in decision making. We uphold the guiding principles for Responsible Tourism which were identified:

Guiding Principles for Economic Responsibility

— Assess economic impacts before developing tourism and exercise preference for those forms of development that benefit local communities and minimise negative impacts on local livelihoods (for example through loss of access to resources), recognising that tourism may not always be the most appropriate form of local economic development.

— Maximise local economic benefits by increasing linkages and reducing leakages, by ensuring that communities are involved in, and benefit from, tourism. Wherever possible use tourism to assist in poverty reduction by adopting pro-poor strategies.
— Develop quality products that reflect, complement, and enhance the destination.
— Market tourism in ways which reflect the natural, cultural and social integrity of the destination, and which encourage appropriate forms of tourism.
— Adopt equitable business practises, pay and charge fair prices, and build partnerships in ways in which risk is minimised and shared, and recruit and employ staff recognising international labour standards.
— Provide appropriate and sufficient support to small, medium and micro enterprises to ensure tourism-related enterprises thrive and are sustainable.

Guiding Principles for Social Responsibility

— Actively involve the local community in planning and decision-making and provide capacity building to make this a reality.
— Assess social impacts throughout the life cycle of the operation – including the planning and design phases of projects—in order to minimise negative impacts and maximise positive ones.
— Endeavour to make tourism an inclusive social experience and to ensure that there is access for all, in particular vulnerable and disadvantaged communities and individuals.
— Combat the sexual exploitation of human beings, particularly the exploitation of children.
— Be sensitive to the host culture, maintaining and encouraging social and cultural diversity.
— Endeavour to ensure that tourism contributes to improvements in health and education.

Guiding Principles for Environmental Responsibility

— Assess environmental impacts throughout the life cycle of tourist establishments and operations – including the planning and design phase—and ensure that negative impacts are reduced to the minimum and maximising positive ones.

— Use resources sustainably, and reduce waste and over-consumption.

— Manage natural diversity sustainably, and where appropriate restore it; and consider the volume and type of tourism that the environment can support, and respect the integrity of vulnerable ecosystems and protected areas.

— Promote education and awareness for sustainable development - for all stakeholders.

— Raise the capacity of all stakeholders and ensure that best practice is followed, for this purpose consult with environmental and conservation experts.

We recognise that this list is not exhaustive and that multi-stakeholder groups in diverse destinations should adapt these principles to reflect their own culture and environment.

Responsible tourism seeks to maximise positive impacts and to minimise negative ones. Compliance with all relevant international and national standards, laws and regulations is assumed. Responsibility, and the market advantage that can go with it, is about doing more than the minimum.

We recognise that the transparent and auditable reporting of progress towards achieving responsible tourism targets and benchmarking, is essential to the integrity and credibility of our work, to the ability of all stakeholders to assess progress, and to enable consumers to exercise effective choice.

We commit to making our contribution to move towards a more balanced relationship between hosts and guests in destinations, and to create better places for local communities and indigenous peoples; and recognising that this can only be achieved by government, local communities and business cooperating on practical initiatives in destinations.

We call upon tourism enterprises and trade associations in originating markets and in destinations to adopt a responsible approach, to commit to specific responsible practises, and to report progress in a transparent and auditable way, and where appropriate to use this for market advantage. Corporate businesses can assist by providing markets, capacity building, mentoring and micro-financing support for small, medium and micro enterprises.

In order to implement the guiding principles for economic, social and environmental responsibility, it is necessary to use a portfolio of tools, which will include regulations, incentives, and multi-stakeholder participatory strategies. Changes in the market encouraged by consumer campaigns and new marketing initiatives also contribute to market driven change.

Local authorities have a central role to play in achieving responsible tourism through commitment to supportive policy frameworks and adequate funding. We call upon local authorities and tourism administrations to develop—through multi-stakeholder processes—destination management strategies and responsible tourism guidelines to create better places for host communities and the tourists who visit. Local Agenda 21 programs, with their participatory and monitoring processes, are particularly useful.

We call upon the media to exercise responsibility in the way in which they portray tourism destinations, to avoid raising false expectations and to provide balanced and fair reporting.

We all have a responsibility to make a difference by the way we act.

We appreciate Calvia's (Spain) intention to host the second conference on Responsible Tourism in Destinations in May 2004; and recognise that this provides a focus for our work over the next two years. The Calvia Conference will focus on the roles of local authorities and tour operators in working to achieve responsible tourism.

We commit ourselves to work with others to take responsibility for achieving the economic, social and environmental components of responsible and sustainable tourism. The Calvia Conference will provide us with the opportunity to share further experience and to assess progress towards achieving better places for hosts and guests. We look forward to hearing about progress in South Africa in implementing its national responsible tourism policy and reviewing developments in other destinations.

References

Allcock, A., B. Jones, S. Lane, and J. Grant. (1994). National *Ecotourism Strategy.* Canberra: Commonwealth Department of Tourism.

Brandon, K. (1996). *Ecotourism and conservation: A review of key issues.* World Bank Environmental Department Paper No. 033. Washington D.C.: The World Bank.

Scheyvens, R., (1999), "Ecotourism and the empowerment of local communities". *Tourism management,* 20: 245-249.

Shiva, V. et al. (1991). *Biodiversity: Social and Ecological Perspectives.* World Rainforest Movement, Penang.

6

Ecotourism and Biodiversity Conservation

Biodiversity is vital to our very existence, and the implications of this alarming loss of biological resources and systems are wide and varied. Ecosystems provide humanity with an absolutely indispensable array of services,maintaining the health of our economy, our environment, our society, and ourselves. 'Biodiversity is our living heritage, providing us with food, clothing, housing, clean air and water, inspiration and spiritual renewal.' The mass extinction of species, if allowed to persist, would constitute a problem with far more enduring impact than any other environmental problem. Biodiversity loss is not only an environmental issue. Many aspects of the economy and social fabric are fundamentally dependant on biological resources. One increasingly important example is that biodiversity is considered to provide the State with a distinct advantage for tourism.

According to evidence from mass extinctions in the prehistoric past, evolutionary processes would not generate a replacement stock of species within less than several million years. What we do within the next few decades will determine the long-term future of a vital feature of the biosphere, its abundance and diversity of species.The conservation of biological diversity provides significant cultural, economic, educational, environmental, scientific and social benefits.

Biological diversity or biodiversity has sometimes been interpreted as limited only to wildlife. However, such is not the case. It encompasses all life on earth. Biodiversity means that variety of all living organisms, including diversity within and between both wild and domesticated species. It also includes the variety within the ecosystems in which these species live. Increasing awareness of the importance of biodiversity and concerns regarding significant losses in global biodiversity encouraged many countries to become parties to the United Nations Convention on Biological Diversity. This convention has three main objectives.

— Conservation of biodiversity

— Sustainable use of biological resources

— Fair and equitable sharing of benefits resulting from use of genetic resources

This Convention is the basis for many initiatives that are significant to various sectors of the economy, including those for agriculture & agrifood, forestry and bioresource sectors. It also includes:

— At the international level, its Biosafety Protocol aims to protect the environment by regulating the transborder movement of living modified organisms.

— At the domestic level, it aims to protect and recover wildlife species at risk of extinction through stewardship, incentives and prohibitions.

Biodiversity includes the variety of all life forms on earth, that is, plants, animals and microorganisms. It deals with the variety of species and not the quantity that is within each of the species. So far, over 1.4 million biodiversity species have been identified and described. It is estimated that there are some 10 to 100 million species in the world and that many species are destroyed every day due to human activity. It should be noted that we have not inherited the earth from our parents, but we have borrowed it from our children. However, we are destroying it every day through overproduction, over-consumption and over-pollution. By progressing on this path, there will be no legacy left for us to transfer to our children. We will be environmentally bankrupt. Environmental damages are continuing at an unsustainably rapid rate.

We have the moral and ethical responsibility to protect, conserve and preserve biodiversity. Philosophically, it can be justified on two bases:

(i) ecosystems and species have the right to exist; and

(ii) on the basis of religious and cultural values, we do not have the right to destroy any species.

These arguments represent the moral intrinsic value of biodiversity that is unrelated to the monetary value of human needs it serves. From the economic viewpoint, many developing countries are cutting trees at an alarmingly high rate. This is justified in the name of economic development of poorer societies, although they are well aware of the long-term impact of this activity on their own welfare and the country's environmental stability.

Unfortunately, preservation of the environment is often in conflict with economic development that affects jobs and social welfare. That is why most of the efforts are made to increase values that are oriented towards human benefits. However, as the population rises, the economic activities become more intensified, thereby creating unsustainable imbalance between consumption and protection of the biosystem.

Values of Biodiversity

Biodiversity is essential to preserving ecosystems and their equilibrium in nature. Some species are not able to compete, while others are able to absorb extra nitrogen. Thus, biodiversity is needed to maintain ecosystems. It is also important in the food chain. At the genetic level, a pool of species should be kept in reserve for future generations. The genetic materials contained in plants, animals and microorganisms have a good potential for agriculture, health and welfare, and environmental purposes. Species can have value as commodities and amenities and they can have moral value, as noted hereunder.

— Value of Species

— Value Product

— Commodity value or direct value

If it can be made into products that can be bought or sold in the market place. For example: Alligators to make shoes, bags; wool from vicuña to make fabric Amenity Value If its existence improves our lives in some non-material way. For example: hiking, bird watching, ecotourism, pet animals.

The decline of biodiversity is the damage caused by human activities, which represents a threat to human development. Human activities that alter or destroy natural ecosystems are likely to cause deterioration of ecological services, which reduces its economic benefits to society. Over the past, the value of the earth's life support systems have been ignored until their loss highlighted their value. A large biodiversity ensures ecological stability, but this balance of nature is now upset due to damage caused to ecosystems by human activities. Erosion of biodiversity is a real tragedy, because it has no restrictions or ownership rights.

Over-utilization induces destruction of this valuable resource. However, biological resources constitute a capital asset with a strong potential to produce sustainable benefits, only if they are used in a sustainable manner. Thus, it is possible to achieve a "win- win" situation between environment and human development, that is, population growth, economic and industry. Unfortunately, most of the benefits due to natural processes are not traded in the economic market. In fact, a renewable natural resource has no value in itself. Usually, such value is assigned by the owners and users in terms of direct and indirect monetary benefits they receive.

The benefits of conserving biodiversity have been categorized into two different classes—local and global. The local benefits refer to many sustainable uses of habitat that will benefit local people. The global benefits refer to the world as a whole that include reduced carbon dioxide from the atmosphere. There are two ways to promote and maintain conservation of biodiversity:

(i) To conserve/preserve biodiversity is by placing limits on the use of habitats, although increasingly more space (land area) will be required to meet the challenges of population growth and economic development.

(ii) To encourage sustainable use of biodiversity, natural resources might be exposed to some abuse due to the level of utilization needed in some areas of the biosystem.

The total economic value of biodiversity is represented by the "use" and "non- use" values. The use values comprise direct and indirect values, while the non-use values represent the intrinsic value.

— *Direct Use Values*: Supply local and global markets with plants and genetic material, medicinal plants and minor products (like nuts, fruits and ecotourism).

— *Indirect Use Values*: Trees absorbing carbon dioxide, wetlands as storm protection, flood protection and water purification.

— *Non-Use Values*: It can be estimated by the debt- for-nature swap, by which an interested conservation agency buys some of the developing country's international debt in return for a promise from the indebted country that it will look after a conservation/protected area.

A per hectare use value can be calculated by dividing the amount received and the area conserved. A non- use value for wilderness an also be calculated by receiving donations to conserve some of these areas, and dividing the amount of donations by the area conserved. The opportunity cost of biodiversity conservation is determined by comparison of costs and benefits of alternative uses of land for other purposes. One of the problems is that biodiversity conservation is not subsidized, but the alternative land uses are. Therefore, biodiversity conservation depends on financial contributions from private individuals, which is unfair. This is one of the main reasons for the high decreasing rate of biological diversity, because, in most cases, private donations are not sufficient to provide adequate protection.

One way to support preservation of biodiversity is that countries that have biodiversity should retain the benefits they are providing to the rest of the world. If they do trade these benefits, they should do so on a sustainable basis and within their protection of biodiversity and environmental regime.

Biodiversity is being severely threatened in all parts of the world, necessitating a truly global approach to its conservation. In a significant step forward in international environmental law, the International Convention on Biological Diversity, which came into force in 1993, explicitly integrates the two objectives of conservation and the sustainable use of biological resources.The objectives of the Biodiversity Convention are 'the conservation of biological diversity, the sustainable use of its components and the fair and equitable sharing of the benefits arising out of the utilisation of genetic resources'.

The importance of this integrated approach to biodiversity conservation was further emphasised at an international level by the Rio Declaration; Agenda 21: Earth's Action Plan.Parties to these two conventions are required to integrate conservation and sustainable use of biodiversity through 'the implementation of national strategies, plans and programs for sectors such as agriculture, fisheries and forestry and for cross-sectoral matters such as land use planning and decision making.'

One important provision in the Convention on Biological Diversity which acts to facilitate this integration of preservation and utilisation of biological resources is the agreement that 'States have the sovereign rights to exploit their own biological resources pursuant to their environmental policies.The emphasis on integration of biodiversity protection measures and biodiversity utilisation practices in this context is not an endorsement of the rationale that biodiversity should only be preserved to the extent that it can be used, and is therefore economically beneficial. Rather, that planning and activities to promote biodiversity protection must be closely integrated with the human processes which act to threaten biodiversity, i.e. the utilisation of biological resources. Planning for sustainable use of these resources is therefore of paramount and international importance in the protection of the diversity of the worlds natural heritage.

Another important international convention which responds to worldwide concern over the fate of the Earth's cultural and natural heritage is the Convention for the Protection of the World's Cultural and Natural Heritage, or the World Heritage Convention. While there is no prescriptive management standards for World Heritage sites, States which have signed the convention recognise that the sites located in their territory constitute a world heritage 'for whose protection it is the duty of the international community as a whole to cooperate.'

The International Union for the Conservation of Nature and Natural Resources (IUCN) (Now the World Conservation Union) is an important international organisation which provides an internationally recognised framework for classifying protected areas based on their management status. All protected areas can be classified using this system which ranges from class 1 fully protected areas, which are managed under strict protection mainly for science or wilderness protection, to class VI protected areas, which are managed mainly for the sustainable use of natural ecosystems. Another more recent international initiative to aid the conservation of biodiversity has been instigated among the world's scientific community.

The problem of loss of biodiversity is not simply an environmental problem, it includes economic, cultural and societal issues, thus the classification of management areas on environmental grounds alone is incompatible with the goal of integrating the management of all of these aspects. Therefore, it is felt that bioregional planning for biodiversity conservation should be redefined to reflect the spatial arrangements of landuse and other cultural elements of the environment: Bioregionalism should become bioculturalism. Biocultural regions need to be flexible, and reflect that there are no specific boundaries to ecological systems, or indigenous cultural systems based on tribal territories, while specific legislative boundaries do exist in the case of local government jurisdiction and modern patterns of land tenure and usage. Any regionalisations therefore need to be treated as flexible and potentially variable, to be used for some purposes, but not others.

It has been recognised and supported at international, national, and state levels, that the successful integration of biodiversity conservation with sustainable economic development of biological resources will be vital in the sustainable conservation of the biodiversity. This is a complex goal in itself, but one that is intrinsic to the achievement of a sustainable relationship between people, economy, and the environment.

It may seem strange to attempt to conserve biological resources by finding new ways to use them, but on the other hand, people and the economic frameworks that guide them tend to value the parts of the environment that they derive economic gain from, or otherwise use. So if sustainable uses can be found for biological resources, particularly economically beneficial sustainable uses, it follows that those resources should be better conserved.

While in some cases commodifying aspects of the environment can prompt increased natural area conservation and promote increased economic independence of regions, in many cases, particularly where there is a lack of legislative control and/or ethical consideration associated natural resource exploitation, the reverse situation can arise. Historically the latter has been the rule rather than the exception; the exploitation or utilisation of natural resources under developed economies has very rarely seen the conservation of those resources. The resources have been unsustainably depleted as a result of uncontrolled and in many cases unethical exploitation of nature, and commonly also culture.

Environmental Impact Assessment (EIA) for development proposals is one tool that may be used by the government to integrate conservation objectives into natural resource based industries from the very start of their operations. EIA assessment can be a useful tool for screening development proposals for compatibility with the goal of biodiversity conservation, however critics argue that insufficient attention is paid to ecosystem protection and preservation in EIA processes, and frequently question the reliability of EIAs to predict the full environmental impacts of development, especially where cumulative impacts and interactions of projects occur.

'A project-by-project assessment can easily fail to predict indirect impacts caused by other sectors.' This is especially important for the sustainable management of natural resources, since sustainability depends on remaining within prescribed limits of resource tolerance and standards of ecological integrity. 'Overall, therefore, EIA processes are valuable and perhaps necessary in conserving biodiversity, but far from sufficient.'

The emerging industry of 'bioprospecting,' the search for new chemicals or genetic material in living things that will have some medical or commercial use,is an example of a new use of biodiversity which could lead to it's conservation. The activity of bioprospecting in some areas, particularly developing economies, is also seen as an opportunity for the sustainable generation of income and funding for purposes such as conservation and health care. Pharmaceutical companies and agribusiness involved in bioprospecting use indigenous knowledge about the properties of native plants as a precursor to screening, and this is happening with little regard for the protection of indigenous intellectual property, and with no equitable sharing of profits.

The tourism industry is a large and fastly growing industry and increasingly becoming a major user of biological resources.Therefore, strategically integrating biodiversity conservation with the future needs of the nature-based and ecotourism industry will be vital in achieving the sustainability of this sector.

From a purely conservation-economics perspective, the main goal of nature-based and ecotourism on protected lands and national parks is to recoup the management costs of those areas from tourism dollars earned.

All of tourism must be ecologically sustainable, and that, for this to be successful, it must contribute to the long-term maintenance of ecosystems and species. Simply to 'minimise impacts' is not enough since, with tourism growth, this will result in incremental damage and inevitable environmental deterioration. There can be no environmental deterioration if the industry is to be ecologically sustainable.

Required accreditation of tourism operations is a powerful tool at the disposal of the government for promoting sustainability among this industry. The 'Green Globe' Accreditation system is perhaps the most widely recognised and credible accreditation system for nature based and ecotourism, ideally however, all tourism activities should be accredited for their efforts to minimise the ecological impact of their operations.

It is important that the Governments have a clear and well defined strategic plan for it's response to the biodiversity crisis in the state. The development of the Biodiversity Strategy needs to become a priority for government, and this can be seen as an opportunity for the development of an innovative and powerful strategic direction for biodiversity conservation and the management of industries which utilise biological resources. It is vital that the future Biodiversity Strategy has a strong legislative support framework that allows for substantial powers of prosecution for environmental offences committed by individuals as well as organisations.

Ecotourism and Biodiversity Conservation Planning

According to the World Tourism Organization (WTO), world tourism in the year 2000, spurred on by a strong global economy and special events held to commemorate the new millennium, grew by an estimated 7.4 % - its highest annual growth rate in nearly a decade and almost double the increase of 1999. Nearly 50 million more international trips were made in 2000 - bringing the total number of international arrivals to a record 698 million.

The number of domestic tourists is still difficult to accurately quantify, but is estimated by some researchers to be as much as 10 times the number of international tourists. Clearly, tourism has a paramount economic role for countries around the world and, if planned and managed correctly, can significantly contribute to sustainable socio-economic development and environmental conservation.

However, inappropriate tourism developments—based mainly on the model of mainstream or mass tourism—are producing severe negative impacts on the natural and cultural environment, including biodiversity. If uncontrolled mass tourism is allowed to continue overrunning many areas of natural and cultural significance, irreversible damage will occur in these areas, which are the repositories of biological and cultural diversity in the planet as well as important sources of income and well-being for all countries and many local communities.

Even tourism developments in urban settings, far away from natural areas, may have unanticipated effects on surrounding lands and waters and the atmosphere, thus affecting biodiversity in many ways. Consequently, the appropriate interaction between biodiversity conservation planning and tourism planning and development has become a key concern for many institutions at the local, national and international levels.

The UNDP/UNEP/GEF Biodiversity Planning Support Programme (BPSP) has a mandate to provide assistance to national biodiversity conservation planners as they develop and implement their national biodiversity strategies and action plans. The integration of biodiversity into other sectors of the national economy and civil society has been identified as a critical indicator of successful implementation of sustainable development practices and of the objectives of the Convention on Biological Diversity (CBD).

To achieve this, UNEP has commissioned a series of thematic studies, each focused on one aspect of sectoral integration. One of these thematic studies is Integration of Biodiversity into the National Tourism Sector. Sustainable tourism has been highlighted recently as an area of major concern both within UNEP and the CBD.

The United Nations Environment Programme (UNEP) Division of Technology, Industry and Economics has, over the past two years, surveyed all main guidelines on sustainable tourism that are already available, and has consolidated and summarised these into a single set of proposed "Principles for Implementation of Sustainable Tourism". Outside of the mechanisms of UNEP and the CBD, a large number of other initiatives linking biodiversity and tourism (including ecotourism) have been undertaken by many organisations, ranging from the World Tourism Organisation (WTO), UNESCO, a number of NGOs, as well as numerous national and regional level destinations, and private tourism companies. It is extremely difficult for national biodiversity planners to (i) identify, (ii) access and (iii) assimilate all of this information on the interface between biodiversity and tourism. It is also difficult for biodiversity planners to see the relevance of much "high-level policy-speak" to their day-to-day activities on the ground. Sustainable tourism has the capability of being a feasible tool for biodiversity conservation by providing economic alternatives for communities to engage in other than destructive livelihood activities, creating new revenue streams to support conservation through user fee systems and other mechanisms, and building constituencies that support conservation priorities by exposing tourists, communities, and governments to the value of protecting unique natural ecosystems.

National and Regional Planning Strategy

Every national and regional planning strategy should strive to improve the social and economic levels of human communities while maintaining environmental integrity in perpetuity. In other words, planning should contribute in an important way to sustainable development, which comprises all human activities. For this reason we should ensure that tourism, which has attained a major socio-economic role around the world, is balanced with other economic, social and environmental objectives.

A national tourism strategy should be firmly based on profound knowledge and wise use of environmental resources, which includes biodiversity. This will be achieved only if we foster harmonisation among all public policies, including tourism and biodiversity conservation planning. This harmonisation must occur at the national level, as well as at the regional (sub-national) and local levels.

Both tourism planning and biodiversity conservation planning lend themselves to a bio-regional approach. It is important to promote valuing of natural resources and tourism assets for inclusion in national accounts. The importance of tourism has to be promoted at a national scale.

One of the challenges that governmental tourism agencies face is that usually they are low priority compared to other government bodies. Tourism agencies need to do a much better job of understanding and demonstrating the importance of tourism to national and regional economies. "Use-it-or-lose-it" conservation strategies aim to achieve their goals through strictly controlled access to ecologically significant areas and resources - and tourism is a key component in this approach, giving value to the natural resources, including biodiversity.

Since there is the need for communities to appropriately value their resources, local education in the environmental and tourism fields are urgently required. It is very important to recognise the need for a symbiotic relationship between biodiversity conservation planning and tourism planning.

Most countries around the world still do not dedicate enough effort towards the conservation of biodiversity, perhaps because they haven't recognised the value and usefulness of this effort. Keeping the biodiversity resource base means enhancing the tourism attractiveness of a country. Unfortunately, in many developing countries, any change proposed in favour of an improvement in the conservation of biodiversity is interpreted as an obstacle for development.

The sustainable positive impact of biodiversity within a comprehensive regional development strategy that considers overall qualitative benefits and not merely quantitative economic gains within a partial sectoral focus is still not understood by the population at large, nor by many politicians. It is important to recognise that many rural communities located in some of the most attractive ecotourism destinations are characterised by extreme poverty.

Many government development plans don't consider their involvement in tourism activities, especially ecotourism, as an option to alleviate poverty. Thus, it is very important to insist on promoting ecotourism to governments as an important tool for rural poverty alleviation programmes. There is a generalised lack of information on the environmental impact of the different economic activities on biodiversity and of mechanisms to evaluate and monitor this impact.

The key factor for success for Costa Rica's tourism sector is sustainability as much of tourism activity as of natural resources. The State plays the role of coordinating entity and regulator of projects and programmes that provide incentives to the communities, and also promotes and generates a real need for a sustainable model as part of environmental, business and local participation schemes.

Tourism activity is one of the most important ways of valuing biodiversity and backing its conservation. National and foreign tourists are counted among the National System of Conservation Areas (SINAC's) main clients. It is they who currently generate the greatest amount of resources for the institution.

With the aim of using tourism as a positive tool for the management of the protected wilderness areas (PWAs), SINAC has defined as its general policy facilitating sustainable tourism development based on responsible practices of planning and management that are in accord with actions for conserving the country's natural and cultural heritage.

Tourism in the PWAs seeks to prevent environmental damage, foster the satisfaction of visitors, provide support to the monitoring of sustainable tourism and is interested in contributing to the country's local economies. This is a good example of a clear objective in planning to use tourism as a tool for natural resource management.

In South Africa, taxing pollution and subsidising products and activities with less environmental damage (including ecotourism), as well as altering interest rates to encourage certain land use patterns are economic mechanisms that are efficiently being used by the government.

In Mexico, the Ministry of Tourism (SECTUR), in coordination with the Ministry of the Environment (previously SEMARNAP, now SEMARNAT), the Commission for the Knowledge and Use of Biodiversity (CONABIO) and several other institutions from the public, private, social and academic sectors, published in the year 2000 a National Policy and Strategy for Sustainable Tourism, which represents a good inter-sectoral effort and contains valuable guidelines and action plans.

In UK, a good example of striving to alleviate poverty is the British government's Department for International Development (DFID's) pro-poor concerns and their interest in funding tourism. Establish a national tourism strategy that prominently includes guidelines for biodiversity conservation planning. For conserving biodiversity through sustainable tourism:

— provide a strong scientific basis,
— adopt an integrated management approach that covers all socio-economic aspects of a region, including tourism,
— carry out ongoing monitoring of environmental impacts, through effective application of Environmental Impact Assessment (EIA), which should be carried out by inter-sectoral technical entities,
— apply the precautionary principle.

Promote and strengthen the decision making process and the standardisation (norms) process in a participatory manner, especially at the local level. Build up institutional capacity of the environmental authorities for follow up of environmental impact assessment and application of prevailing norms. Apply integrated land use planning at a regional scale, taking into consideration the local communities' opinions.

Within the framework of the national tourism strategy, establish a sound ecotourism policy of an inter-sectoral nature that will provide viable options for biodiversity conservation and sustainable development especially at the local rural level.

Concentrate efforts on mid and long-term development policies without being pressured by periodical changes in government administration. Foster the use of taxes, subsidies and interest rates to reduce the negative environmental impacts of economic activity and enhance the positive effects.

Making loans more easily available is a key point, in which the government should also participate.

Both biodiversity conservation planning and tourism planning and development are complex, interdisciplinary and inter-sectoral phenomena, thus appropriate interaction and integration is even more so challenging. There is an overall need to develop effective inter-sectoral mechanisms that will ensure the harmonious interaction among all stakeholders and a symbiotic linkage between biodiversity conservation planning and tourism planning and development.

For this reason, practical and dynamic coordination bodies (whether they be called commissions, committees, councils or agencies) for ensuring this appropriate interaction must be set up. These bodies should include genuine, fully empowered representatives of the different stakeholders or sectors: government, private tourism sector, NGOs, local communities, and financial institutions.

Although it is difficult to have the tourists themselves in a committee of this sort, their input should be sought. Inter-sectoral cooperation for the appropriate, sustainable and interactive development of tourism and biodiversity conservation planning must occur at the local, national, and international levels. Jurisdictions and responsibilities of the different agencies, authorities and organisations should be clearly defined and complement each other.

Government representatives in these inter-sectoral coordination committees should at least come from the ministries (or equivalent government or para-statal agencies) dealing with tourism and the environment, and it is convenient also that the ministries dealing with agriculture, fisheries, and education be included. Sometimes it is not necessary to create a new entity.

A preexisting body may be strengthened so as to develop a mandate of cooperation and non-duplication - one which is legislated, if need be. In a number of countries, these inter-sectoral coordination bodies have been established, with varying degrees of success. In every case it is important to first create the appropriate political environment between government departments (tourism, environment, agriculture) before bringing the other sectors (NGOs, tour operators, local communities) together.

The role of NGOs is vital, and should include local, national and international NGOs. Bilateral and multilateral development aid agencies (where appropriate) should collaborate, but always serving the local and national interests of the corresponding country. There must also be environmental education for those who are taking decisions, both in the fields of conservation and tourism.

In Australia, CRC is a network of Cooperative Research Centres in Australia, one group of which is dedicated to sustainable tourism. The Sustainable Tourism CRC is a very effective model of how universities can work with government and the tourism sector to develop and implement a research agenda that works for all three groups.

Housed in universities throughout Australia, the research is undertaken by academics but directly responds to the needs of government and industry. It helps all facets of the sustainable tourism industry to become linked. In Chile, in 1994, an entity called the National Commission for the Environment (CONAMA) coordinated through the President's Secretariat was created.

In Costa Rica, this country has successfully integrated the management of national parks, wildlife and forests into a single organisation, the MINAE's (Ministry of the Environment and Energy) SINAC (National System of Conservation Areas). The system is made up of eleven Conservation Areas that comprise 25% of the entire country, providing a good scenario for inter-sectoral synergy. Strengthening of inter-sectoral coordination is achieved through clear mechanisms for participation of the private and state sectors for planning tourism: e.g., the National Accreditation Commission for the Certification for Sustainable Tourism (CST).

In Costa Rica, communities have formed associations and cooperatives in different places throughout the country that have, as their main activity, local tourism. Private reserves, which at the moment total more than 100, and cover 1% of the national territory, rely on ecotourism as their main source of revenue. They carry out their activities in coordination with, and backing from, the Costa Rican Tourism Institute (ICT) and the MINAE, in the case of ecotourism, and in the case of agrotourism projects, with the ICT and Agrarian Development Institute (IDA).

In Costa Rica, the Bandera Azul Ecológica ("Blue Ecological Flag") programme was established by ICT in 1996 by Executive Decree as a coordinated action carried out by MINAE, AyA (Institute of Water and Sewage), CANATUR (National Chamber of Tourism), and the Ministry of Health. This programme awards the Blue Flag to Costa Rican beaches that comply with requirements of cleanliness, environmental education, and community organisation, thus warranting the sanitary and aesthetic qualities of the beaches.

The Belize Tourism Industry Association (BTIA) was formed in 1985 to bring together tourism-related interests to meet the challenges of the industry and act as an important link between the public and private sectors. BTIA has chapters in each district of Belize, and includes a broad range of individual members from all areas of the industry as well as representation from associations such as the Belize Hotels Association, Belize Tour Guides Association, Belize Tour Operators Association, Belize Ecotourism Association and others.

A perceived weakness of BTIA is that it functions more like a membership-based organisation rather than an umbrella organisation. As the latter, it would be able to function more directly as an advocacy for the industry. In Canada, there are a number of senior level government committees which are designed to enable united action across the country, in the various jurisdictions which have independent authority over resources.

In Canada, the Canadian Tourism Commission (CTC) developed their vision and mission tapping into the direction of a 20-member team of industry experts, then presented them to industry stakeholders, provincial and territorial governments, and destination marketing organisations, to get consensus on a shared vision and mission, and to encourage a unified Canada perspective.

In Canada, the Biosphere Reserve Ecotourism Initiative is one of a number of tourism partnerships in the CTC Product Club Programme. The Ecotourism Product Club is a partnering programme between individual Canadian Biosphere Reserves, local communities and the private sector. The goals of this partnership are to: encourage and package sustainable tourism opportunities based in nearby communities and tapping into the resources of the Biosphere Reserves; and help communities value protected areas through demonstrating that economic benefits can emerge from ecotourism.

Strengthen the coordination of tourism and biodiversity conservation planning government policies at both the national and local levels. Foster the participation of biodiversity conservation planners in meetings dealing with tourism planning and, conversely, the participation of tourism planners in discussions of biodiversity conservation issues.

Tourism needs to be involved in biodiversity and biodiversity in tourism. Create dynamic and practical inter-sectoral mechanisms (e.g. committees) for effectively coordinating interaction between biodiversity conservation planning and tourism planning and development. Ensure that these committees have representatives not only from the government sector, but also from the tourism industry, NGOs, local communities and the universities.

Nations around the world have growing economic needs and generally growing populations. However, their land area normally remains the same and this produces many conflicts regarding use of the land and its resources. Utilisation of land for tourism is becoming an increasing trend in many countries. It is important to ensure that national development plans contain a set of development guidelines for the sustainable use of land, water and natural resources.

The development of a diverse tourism base that is appropriately integrated with other local economic activities should be promoted. The only type of tourism which should be permitted in ecologically relevant and fragile natural areas is ecotourism. Ecotourism provides viable economic alternatives for other activities which are less sustainable—and many times downright destructive to the environment—, such as uncontrolled logging, extensive animal husbandry, mass tourism or mining.

This may bring about conflicting interests of water supply and other vital resources between tourists, the needs of the local population, and intensive agriculture. For this reason effective national zoning plans for land use are urgently required (and of course, effective enforcement of those plans). Ecotourism should not be seen as a panacea or as a monoculture in a given area, but should be considered a complement for other sustainable economic activities that make wise use of the natural resources.

All relevant stakeholders should be involved in the development of sound management plans for land-use. Means for providing the organisation, facilities and enforcement capacity required for effective implementation of those management plans should be put in place.

In Australia, in the Shark Bay Region of Western Australia, the 1995 Shark Bay Regional Plan coordinated a World Heritage Strategic Plan, several tourism strategies and several fisheries management strategies. In order to effectively manage future development and use, the Department of Conservation and the Ministry of Planning worked together to ensure that their planning was complementary and there were no areas of conflict as regards use of land and coastal zones.

In Costa Rica, the Conservation Area is an original national regionalisation structure created by MINAE (Ministry of Environment and Energy) and SINAC (Conservation Areas National System). A Conservation Area is defined as that territorial unit governed by one and the same strategy for development and administration, where there is interaction between private activities (including tourism) and state activities on issues of management and conservation of natural resources, and where sustainable development solutions are pursued jointly with civil society.

In this way, SINAC's administration covers all of the national territory (including protected areas) where exploitation of biodiversity activities and natural resources in general are promoted and regulated. Each area has a set number of protected wilderness areas (PWAs), including buffer zones. There is coordination between state institutions and civil society that carry out activities in the areas.

In Costa Rica, as for protected areas, SINAC in coordination with NGOs, has widely invested in the promotion of the different and numerous PWAs in the country and in bettering the infrastructure for attention to the public. This effort has had its positive impact, as was shown by a recent survey on tourists ' perception on international tourism but more so and especially on national tourism, as reflected by the growing number of visitors to the PWAs.

In Costa Rica, both in conservation areas and protected areas the ENB (National Biodiversity Strategy) establishes a line of action and top priority activities according to five-year planning periods, where the tourism sector is a key factor. In the Strategy, ecotourism with ample participation by civil society and in coordination with the government, is visualised as the top priority tourist activity to be developed and strengthened for conservation and sustainable use of biodiversity.

Ensure that tourism planning is undertaken as part of the overall development plan for any area (including its land-use plan), and not undertaken in isolation. Focus on ways in which different interests can complement each other within a balanced programme for sustainable development, enabling the different stakeholders to work alongside each other.

Minimise negative impacts of tourism on the natural and cultural environment and promote tourism as a tool to protect important natural habitats and conserve biodiversity in accordance with the Convention on Biological Diversity. Ensure that, wherever tourism occurs or is liable to occur, the corresponding land-use plan should include a component of tourism land-use, carefully zoning the areas as regards the type of tourism that should take place: high, middle or low intensity.

The inclusion of natural areas (and provisions for their conservation) is a vital element of any zoning plan. Ensure that the only type of tourism that will take place in vulnerable and fragile natural ecosystems follows the principles of ecotourism, considering it a viable option for minimising negative impacts and promoting positive environmental and socio-economic contributions.

Ecotourism should always be carried out with the active involvement of the local communities. Foster the creation of links between natural protected areas and other ecotourism destinations by means of biological corridors that will amplify biodiversity conservation to a larger regional level.

Sustainable Tourism Industry

In every case, tourism should pay attention to the "triple bottom line": economic, environmental and social factors must be attended to simultaneously. This implies the need for integrated management and the adoption of an ecosystem approach, as advocated by the CBD. Continuous management of tourism is just as important as proper planning and development. It is imperative to provide incentives for the wide range application of environmental management systems.

The only viable relationship between tourism and nature conservation is a symbiotic one. It is not enough to have a situation of coexistence and certainly nobody benefits from a conflictive relationship. Tourism management needs to form part of biodiversity management planning. The allocation of land uses must be carefully coordinated and inappropriate activities that damage ecosystems should be strictly regulated.

This may be done only by strengthening and developing integrated policies and management that cover all socio-economic activities in the different ecosystems, including terrestrial, coastal and marine zones. Management solutions are also needed for simple, but persistent, problems such as litter. It must be emphasised that enjoyment of biodiversity and natural areas is not only for rich foreigners, but for all national inhabitants.

Ecotourism is made up of visitation by both national and international tourists. The former component is usually more sustainable than the latter if a sufficient standard of living exists in the country (i.e. domestic tourists possess the financial means to visit, and consequently support, protected areas). The different sectors must understand the tourism market for cultural and natural heritage products, and how this is linked to tourism's ability to support conservation through product demand.

Understanding the experiences and products tourists are looking for, enables protected area managers to tailor certain aspects of the destination for the desired type of tourist. Accurately forecasting the amount of anticipated visitors enables planners to lobby for and develop sufficient infrastructure.

It is important to demonstrate how the private sector can implement environmental management plans, using low cost methods first, and then use any left over money to retrofit, making the tourism facility more sustainable. It is necessary to show the large hotel chains that environmental management brings a profit. Using environmentally friendly techniques saves money for hotels and all other tourism service providers.

In Australia, a national agenda to support sustainable tourism (including ecotourism) as a tool for conserving biodiversity and for better use of natural areas. The National Tourism Strategy was formulated in 1992 to, among other goals, enhance community awareness of the economic, environmental and cultural significance of tourism.

The Strategy's environmental goal is to provide for sustainable tourism development by encouraging responsible planning and management practices consistent with the conservation of Australia's natural and cultural heritage. In 1994, within the framework of this strategy, the National Ecotourism Strategy was published also by the Department of Tourism.

In the strategy, integrated management of biodiversity resources and sustainable tourism, with ample participation of civil society and in coordination with the government, is considered as the top priority for national development. As part of sustainable tourism, Costa Rica seeks an activity where there is better distribution of resources in the different regions of the country and direct involvement of rural communities, while at the same time minimising negative environmental impacts.

In Cuba, following the 1992 Earth Summit, where Cuba was one of only two countries to obtain the highest rating for implementing sustainable development practices. The government set up a National Programme for Environment and Development and created a series of new institutions to continue along a course of sustainable development. The new institutions include a National Commission on Ecotourism, made up of tourism officials, environmentalists and scientists, created to ensure integrated management of biodiversity resources and tourism activities.

Conventional Mass Tourism

The conventional mass tourism is still the mainstream of the tourism industry and it is quite probable that this situation will prevail for some time. For this reason it is vitally important to aim our attention on mass tourism, striving to apply measures to make it more environmentally friendly and minimising its negative impacts on biodiversity. We should not consider only ecotourism linkages with biodiversity conservation, but also linkages of mass tourism, especially the effects of big hotels on the environment and how their design and operation can become more environmentally friendly.

At a global scale, perhaps providing a number of ecolodges is not going to make much of a difference- ultimately we have to affect the larger tourism industry. This means we have to consider how to improve the environmental record of very different items like airlines, airports, big amusement and theme parks, golf courses, and sports stadia.

Training to develop skills of hotel owners and operators to understand what sustainable tourism is and education about best practices are vital activities. There is a need to strengthen and to revise legislation so that this approach is well understood and widely disseminated.

Environmental legislation should act as a motivation force, and also as a base for certification. Also, a widespread educational campaign so that tourists will be demanding environmentally-friendly hotels is urgently needed. Tourism shouldn't be only market driven. In Africa, for example, people feel bad about tourism use proscribed to the community.

A cause for conflict arises when developing nations are told to be sustainable whereas western countries can have the huge hotels. It is vital to disseminate codes of ethics for conventional tourists, which will serve as a tool for alleviation of negative impacts. The effects of negative impacts are frequently long term and not always obvious in the short term.

Saving water and energy by reducing the number of towels used in hotel rooms has become a cliché - but only because notices have made a difference in hotels around the world. In analysing mass tourism impacts, both new tourism facilities and preexisting tourism facilities must be considered. In the former case, the application of minimal environmental standards for siting of new tourism services and facilities is urgently required.

In the latter case, methods for improving the operation, making it more environmentally-friendly, should be applied, through retrofitting or adding new, more appropriate technologies. In every case, the benefits to the tourism sector (market demand, economics, effective management) must be persuasively demonstrated. It is not a matter of sanctions and pressuring, rather encouraging the tourism sector to become more environmentally friendly (which will result in economic benefits for them).

For example, water heating in many conventional hotels around the world is currently very inefficient and costly, so that wide use of alternative energy sources should be more than welcome by mainstream tourism operations. Also, many traditional beach destinations are experiencing a loss of repeat visitors because of water pollution, so that more environmentally-friendly practices are definitely in the interest of beach resort owners and operators.

Cruise ships cause enormous environmental damage. It is estimated that they discard many thousands of tons of untreated waste into the oceans of the world every day. Strict regulations have to be applied to this type of destructive tourism.

In Australia, for the 2000 Olympic Games the Sydney Organising Committee for the Olympic Games (SOCOG's) mission was "to deliver the most harmonious, athlete oriented, technically excellent and culturally enhancing Olympic Games of the modern era". SOCOG used its best endeavours to set a new standard of environmental excellence for organising and staging a large sporting event. To this end, SOCOG was committed to strict environmental guidelines, guided by the principles of Ecologically Sustainable Development (ESD).

Environment was considered by the organisers as the "third pillar of Olympism". Among the major environmental achievements were: the SOCOG Environment Programme was established in 1996, early enough to allow sufficient time for staff to be integrally involved in planning; carrying out of the Olympic Greenhouse Challenge, a major project to assess the greenhouse impact of the Games for minimising greenhouse gas emissions; a programme of environmental education (including a waste education plan) as a component of staff training; an environmental specification for sponsors, licensees and suppliers; an integrated waste management solution; packaging and foodware specification to control inputs into the waste stream; an Olympic Results Information Service (ORIS), an electronic system which reduced the huge amount of paper required to provide media with results; saving water and energy at all Olympic buildings and facilities; facilitating access of spectators through public transport, resulting in energy conservation and pollution and greenhouse gas avoidance; Olympic merchandise had minimal packaging and minimised the use of PVC.

Encourage collaborative research on the impacts, promotion and incentives of mass tourism. Enforce current environmental laws, regulations and norms on waste management, air pollution and monitoring devices. Overcrowding, misuse of natural resources, polluting of air and water, the construction of infrastructure and facilities, and other activities associated with tourism, all produce impacts on the environment. These impacts may be not only physical, but also cultural.

Negative impacts of tourism vary according to the nature and number of tourists, the type of physical facilities and the way tourism is managed. The individual tourist normally has a relatively small impact. Problems arise, however, if the number of tourists is large or the resource overused.

Thus although tourism can be a lucrative source of revenue for a country, a tour operator or a specific destination, it can also represent a cause of serious damage to the environment, including the biodiversity resources. Tourism impacts on the environment are manifold: impacts on geological exposures, minerals and fossils; on soils; on air and water resources; on vegetation; on animal life; on sanitation; on the cultural environment; and aesthetic impacts on the landscape.

These different impacts, which are actually manifestations of change on the environment, rarely occur singly and their ecological effects are usually very complex. There are different ways of minimising negative impacts of tourism: through widespread environmental education of tourists and the tourism industry, through strict enforcement of laws and regulations, and through effective visitor management.

Visitor management begins before arrival, then occurs on the ground at the site, and finally after the visitors leave through continual communication. The process of environmental impact assessment (EIA) is one of the most effective methods for determining whether a project will be sustainable, and if so, for developing safeguards to ensure its continuing sustainability.

EIA aims to ensure that the likely outcomes of any development (including tourism developments) are addressed at an early stage so that disastrous environmental and social consequences can be avoided. EIA applied to tourism projects should play a crucial role in government-decision making in every country.

Zoning, a vital component of land-use planning, is the process of applying different management objectives and regulations to different parts or zones of a specific area. Zoning is a tool required in any land or water-use plan (including marine, coastal and fresh water areas) and should always be used to regulate utilisation of the land and water resource over the long term.

Zoning should be applied at all the different levels: national, regional and local. In the first case, zoning plans should be carried out by government authorities at the highest (e.g. federal) level. In every case, zoning must be comprehensive, considering the different socio-economic activities (obviously including tourism, when it occurs) , but also natural areas which should be left undeveloped.

Zoning plans should solve conflicts of interest with water supply and other vital resources, including biodiversity. Understand that some areas of relevant biodiversity should be conserved for their own value and that not all relevant biodiversity zones should experience visitation. Biodiversity has is own value. As regards tourism, it is important to have a zoning scheme which should cover the different possible tourism activities. A good example of a comprehensive tourism zoning plan includes the following specific zones:

a) Strictly protected zone, where the presence of all types of tourists and tourist infrastructure are strictly prohibited.

b) Restricted tourism zone, where access is allowed only to a limited number of tourists, usually on foot.

c) Moderate tourism zone, where visitors are encouraged to carry out diverse low -impact activities compatible with the natural and/or cultural environment.

d) Semi-intensive tourism development zone, which should always be an area of limited extent (especially when near environmentally-sensitive natural areas), where some moderate-impact facilities are included (e.g., ecolodge, visitor centre, limited parking areas).

e) Intensive tourism development zone, which should only occur in highly popular mass tourism destinations (e.g. beach resorts, ski resorts, amusement and theme parks), where a considerable degree of concentration of tourists and tourist facilities take place.

Obviously, in ecotourism destinations (especially in protected areas), there is nor room for this zoning category. But, even in the case of mass tourism destination areas, planning should endeavour to minimise negative impacts, including pollution of air, water, and soil resources. The different zones indicate where and what type of physical infrastructure and services should be provided, by means of a clear categorisation of modality and intensity of land use (and use of natural water resources and other natural resources), striving in every case to minimise negative impacts on the natural and cultural environment, as well as optimising the ecotourists' experience.

Zoning also indicates to us where facilities, activities or services should not be developed. In essence, a zoning scheme shows the development suitability of the different portions of a site. Those activities which are carried out in each zone are normally mutually exclusive (and often conflictive), so that zoning decisions must be taken very carefully. Allow diversity through wise zoning of tourism activities, providing tools so that people (both locals and tourists) can decide upon which zone they fit into (or they prefer to visit) and then follow the regulations related to their zone.

Environmental Carrying Capacity

Environmental carrying capacity is the capacity of an ecosystem to support healthy organisms while maintaining its productivity, adaptability, and capability of renewal. Tourism carrying capacity is a specific type of environmental carrying capacity and refers to the carrying capacity of the biophysical and social environment with respect to tourism activity and development.

It represents the maximum level of visitor use and related infrastructure that an area can accommodate. If it is exceeded, deterioration of the area's environmental resources, diminished visitor satisfaction, and/or adverse impacts upon the society, economy and culture of an area can be expected to ensue.

The basic components of tourism carrying capacity are:

— biophysical,

— socio-cultural,

— psychological, and

— managerial.

Over the last decade or so the tourism carrying capacity concept—and several related methodological tools—have been heavily criticised as being oriented excessively towards quantitative considerations. Alternative methodologies, such as LAC (Limits of Acceptable Change) and VIM (Visitor Impact Management) have been developed (especially for relatively undisturbed natural areas. The shift in attention from an appropriate use level to the desired condition is the basis of LAC's revised approach to visitor carrying capacity.

The LAC approach concentrates on establishing measurable limits to human induced changes in the natural and social setting of a specific area, and on identifying appropriate management strategies to maintain and/or restore desired conditions. VIM, developed by the National Parks and Conservation Association of the USA, is a technique for assessing and managing the environmental and 'experiential' impacts of increasing numbers of visitors to natural areas.

VIM recognises that recreational impacts on the environment and the quality of the recreational experience are complex and influenced by factors other than use levels. Traditional carrying capacity methods as well as LAC and VIM techniques are all management tools for minimising negative environmental impacts.

Tourism as a Multi-sectoral Phenomenon

Since tourism is a complex and multi-sectoral phenomenon, it is evident that the local communities should always be considered as a vital sector actively participating in the tourism process of the corresponding region. In every case, it should be the community's own decision to actively participate or not in the tourism process.

Cultural shock (in both directions: from tourist to community and from community to tourist) should be avoided at all costs. This is a very complex issue, which should include the careful consideration of cultural issues, undertaken by social scientists and anthropologists. It is important to recognise that the development impact of tourism will not be uniform: it will vary widely within and between communities.

Enhancing the livelihood impacts of tourism does not mean simply maximising the number of tourists or tourism developments, or maximising wage income; a wide range of costs and benefits need to be taken into account. Benefits should be considered on a long-term perspective, so as to achieve sustainability.

It is as important to address negative impacts as to maximise positive ones and to address impacts on people's assets and existing activities, not just direct contributions to household income and security. In every case, tour operators should make long-term commitments to communities.

Careful planning and design, based on an understanding of local livelihoods, can greatly enhance the positive impacts of tourism. Maximising livelihood benefits needs a good understanding of what people most need and want (their livelihood priorities) and of the complex ways in which tourism options affect livelihoods (direct and indirect livelihood impacts).

This requires a considerable role for local people in decision-making. This can be done either by delegating tourism rights to community level, and helping communities with participatory planning (a "bottom-up" approach); or by ensuring that government planning processes are participatory and responsive to local needs; or by ensuring, through government incentives, that planning by private entrepreneurs is responsive to local needs. The details of how to enhance livelihood impacts are location-specific.

The principles of recognising that a range of livelihood concerns are important, and supporting systems that enable local people's priorities to be incorporated into tourism decisions, can be generally applied. Some study shows that growing numbers of tourists would like more meaningful contact with local communities, including informative interactions. Diversifying to meet this demand could provide low -cost economic opportunities for local people, creating a more rounded and sustainable tourism product.

Envision ecotourism as a motor for local sustainable development and a job generator. It is vital to reach agreements between tour operators and the community so that the community is a member of the organisation and has ownership over the process. The local community must come to value nearby natural areas (including protected areas) before it will protect them. Management and stewardship by the local community is an important part of education and vital to the long-term survival of the resource. It will also contribute to raise pride regarding the local heritage.

In many rural areas around the world, local inhabitants have shifted from being hunters to serving as park rangers or ecotourist guides. When planning an ecotourism development, the first question to the local community should always be: Do you want ecotourism to take place in your area? But do not just ask, provide information on potential negative and positive impacts. Also discuss this in the context of other sectors (e.g. agriculture). Be respectful of their decision.

Ecotourism projects should strengthen community organisation. Always apply participatory planning and self management principles to tourism development. Ensure that traditional local resource managers are given statutory rights over the areas that they manage. People need clear legal rights over the area that they live in before tourism which benefits them and conservation can take place. Encourage capacity building and empowerment for local communities.

Foster links between all-inclusive resorts and local enterprises, e.g. local food suppliers, daily market place, local excursions, etc. Promote symbiotic relationships between big hotels and smaller tourism suppliers, including small lodges. Avoid isolation or enclaves and have tourists be in contact in social and natural environment (when desired by community).

Recur to social scientists and anthropologists (especially those with long local experience) for advice in developing activities and products that will be acceptable and useful to the entire community, e.g., participatory dances and ceremonies, local arts and crafts, regional foods, etc. The product needs to move beyond being a simple merchandise, serving the tourists as being part of their travel experience.

If they are in agreement, foster the active participation of indigenous people in the ecotourism process in such a way that they obtain tangible socio-economic benefits and also, they contribute to the ecotourist's experience. Minimise traveller impact on local cultures by offering appropriate literature, briefings, leading by example, and taking corrective actions.

Develop educational and environmental awareness campaigns and training programmes among the local communities that also include understanding what tourism is about (including positive and negative effects of tourism in local communities). The best ecotourism guides usually turn out to be the local people (with the right training). Give opportunities to the local people for communicating to the tourists their traditional perception of their environment.

Carry out training programmes for tour operators who are working with the local community in order to minimise cultural shock and negative impacts. Diversify tourism experiences. Consider ethnic tourism and agrotourism as excellent complements of ecotourism and conventional tourism. Define whether employment or ownership will give the greater beneficial impact to the community.

Encourage both direct and indirect benefits from tourism, e.g. selling of handicrafts, local language teaching, etc. Encourage and/or facilitate financial support (especially micro-credit schemes) to communities over realistic time periods. Soft money will be a useful tool for economic support but it needs to come with a programme of education based on financial management. Ensure that independent and financially sustainable community businesses are supported by government promotion of tourism. Simplify forms for obtaining micro-credits.

Micro-creditors still need to be paid if the business fails. A safety net needs to be developed. Develop schemes for identifying environmental and social capital (what could be termed 'environmental/social economics'). Transfer consumptive and non-consumptive values to general national accounting. Recognise that tourism is a foreign exchange activity, therefore communities promoting tourism should be eligible for international financing at the same.

In January 2001, the UNEP Tourism Programme produced a final version of Principles for Implementation of Sustainable Tourism for UNEP's

21st Governing Council Meeting. UNEP surveyed a considerable number of the main guidelines that are already available worldwide, and consolidated and summarised these into a single set of principles, in a general and concise way. These principles, which hopefully will be widely applied, are grouped into four main themes:

— Integration of Tourism into Overall Policy for Sustainable Development

— Development of Sustainable Tourism

— Management of Tourism

— Conditions for Success

The World Tourism Organisation has launched the Global Code of Ethics for Tourism, which addresses environmental, social and economical principles, practically defining a baseline for what sustainable tourism should be.

Certification System

Every country needs a national certification system in the field of tourism; it needs to be applied by an inter-sectoral body (that should include NGOs, communities, tour operators, etc.). National governments need not be the certifiers. The certifier is preferably the result of a cooperative effort (e.g. a commission) that involves scientists, communities, industry, etc.

The certification programme must have official recognition by all stakeholders. Certification schemes need to be independent, non-profit, transparent, and credible. If certification is to be used in the field of sustainable tourism it must bring benefits to biodiversity conservation. A standard is a level that every one must meet (e.g. safety standards for being a river raft guide); certification should place more requirements, raising the bar in terms of the criteria that need to be met.

Certification is something destinations must decide for themselves and should be carried out on specific measured environmental performance. Local or regional certification schemes are normally preferable, but based on international guidelines, and they must include biodiversity as one of the main components.

Certification should provide economic benefits to all stakeholders (including tour operators). Most tourism certification in the past has not duly considered factors like environmental quality, biodiversity conservation, and respect for local cultural traditions. In the future, certification should appropriately link tourism with biodiversity conservation, and should be performance-based. Biodiversity (including conservation, maintenance, and enhancement), socio-economic (comprising gender issues , level of participation of local people, and benefit sharing), and cultural criteria must be considered in every certification scheme.

It should be recognised that many certification programmes are currently unreliable (resulting , e.g. in lack of training in hotels). Try to attract tourists who are already complying with certification schemes. In Australia, the National Ecotourism Accreditation Program (NEAP) was launched by the Ecotourism Association of Australia in 1996, and is jointly run with the Australian Tour Operators Network.

The NEAP arose out of the fundamental problem of how to distinguish between genuine ecotourism operators and other operators who operate in natural areas. At the end of 1998, there were about 130 attraction, accommodation and tour products accredited, according to eight basic sustainability principles. Depending on how many points operators achieve, they can either be awarded accreditation or advanced ecotourism accreditation.

The Accreditation Program has been critical in helping ecotourism operators improve the profile of their products, which in turn has led to greater customer recognition and an emerging market edge.

The Mohonk Agreement is an agreed framework and principles for the certification of ecotourism and sustainable tourism, which was unanimously adopted at the conclusion of an international workshop held at Mohonk Mountain House, New Paltz, New York in November 2000. This agreement contains a set of general principles and elements that, according to the workshop participants, should be part of any sound ecotourism and sustainable tourism certification programmes.

Participants came from 20 countries and delegates represented most of the leading global, regional national, and sub-national sustainable tourism and ecotourism certification programs, conservation and environmental organisations, and others with expertise in tourism and ecotourism certification and environmental management. Workshop participants recognised that tourism certification programmes need to be tailored to fit particular geographical reasons and sectors of the tourism industry, but agreed on a basic set of principles that, in their opinion, must frame any ecotourism and sustainable certification programme.

Biodiversity and Nature-based Tourism

Tourism is one of world's fastest growing industries. It is a major source of foreign exchange, domestic product, income and employment. Within this sector, a trend described variously as nature-based tourism and ecotourism (NBE) has emerged as a strong segment over the past few decades. The prospects for expansion of this trade, based on both international and domestic visitors, are significant.

Historically, the increase of NBE worldwide was largely a result of the increase in awareness of the plight of the world's dwindling biological diversity and a reaction against mass tourism and its demonstrative uncaringness of conservation issues. The conservation of biological diversity, usually shortened to biodiversity, is now seen as a priority of national governments and the general community.

Nature-based tourism involves education and interpretation of the natural environment and is managed to be ecologically sustainable. It recognises that 'natural environment' includes cultural components and that 'ecologically sustainable' involves an appropriate return to the local community and long-term conservation of the resource.

The term 'ecotourism' can be defined as travelling to relatively undisturbed or uncontaminated areas with the specific object of studying, admiring, and enjoying the scenery, its wild plants and animals, as well as any existing cultural manifestations found in these areas. There are many other definitions sprinkled throughout the literature and an academic industry which has thrived on the analysis of the 'ecotourism' phenomenon.

In the tourism industry as a whole the terms 'ecotourism' and 'nature-based tourism' are almost always used interchangeably and indiscriminately. We have intentionally avoided adhering strictly to any definitions since we believe that the concept of ecotourism has already matured and developed beyond the original attempts at definition, that the whole industry is changing rapidly, and that elements of nature-based and ecotourism can enter all segments of the tourism market. The future of NBE and biodiversity conservation is dependent on maintaining a flexible approach to their interrelationships.

There will never be a firm division between tourism and ecotourism. Ecotourists must define themselves as an avantgarde camp that brings out the best of the tourism market and provides a model for the rest of the world.

Ecotourism is a situation where 'the idea of a symbiotic relationship between tourism and environment becomes most apparent', and that tourism facilities and services exist in a continuum, from those which are natural resource dependent for their operation, through those which are independent of natural resources. Ecotourism and nature-based tourism can form part of many types of travellers' experiences, varying from a few hours of nature-appreciation, through to intensive long-duration tours of a month or more. That the term ecotourism is clouded in confusion, may well be a reflection of the fact that it has less and less relevance in an industry where the differences between ecotourism and tourism are becoming less distinct. Many industry members including operators, agents, wholesalers and government marketing managers now avoid the term because they feel that it has become misleading, has gained a poor image, is not particularly useful, and means very different things to different people.

Given its tiny base, it is surprising that so many resources have gone into the development of ecotourism strategies, plans, workshops and policies. Logic would dictate that the intention of these strategies is to consider ecotourism as a broader concept than is often stated and assumed, and includes operators and properties across the tourism spectrum, regardless of size.

Responsibility can be and should be held by all facets of the industry. It is the industry as a whole in its policies, plans and negotiations that should be supporting the conservation of biodiversity because, amongst other things, it is demonstrable that when new national parks or reserves are created, the industry as a whole benefits from the opportunities created.

In considering the scope of the potential for integrating NBE and biodiversity conservation, we have considered NBE to embrace the full spectrum of nature and culture-related tourism. This is not limited to tours, but includes walks, talks, resorts, and other accommodation, and other facets of the industry and tourism activities, whether private or industry based, and people partaking of nature-oriented activities also use the same facilities as other tourists.

NBE utilises a diverse range of places. The 'natural environment' encompasses a large part of world and, while much NBE occurs outside protected areas, the large proportion of ecotourism and nature-based tourism occurs on national parks, many within a day's drive or less of major cities. Despite this, NBE does not necessarily occur in areas of high biodiversity value, as a large proportion of national parks have not been reserved for, and are not necessarily of high biodiversity conservation value.

Historically, reasons for reservation have been catchment protection, scenic value, recreational amenity, and landscape protection, amongst others. In more recent decades, increased awareness of the inadequacy of conservation of biodiversity in national parks and other reserves has led to new approaches being developed and in some cases utilised for identification and reservation of national parks and reserves.

The principle of ecological sustainability is often taken to mean little more than minimising impact. This philosophy is based on accepting that there will be an impact from tourism. Incremental small impacts inevitably lead to long-term negative ecological consequences which cannot be sustainable. All tourism, not just ecotourism and nature-based tourism is dependent on the natural resources, and must be a more positive contributor to ecological sustainability, by reversing negative impacts and providing some nett gain through contributions in kind, practice and finance.

The concept of ecotourism has evolved over the past decade from a reaction to mass tourism to a force which is contributing to the general greening of the tourism industry. In this sense it can be seen as a process and its importance in inducing change in the tourism industry may be more significant than its categorisation as a small niche market of small operators. The greening of tourism is essential for the ecological and sociological advancement and sustainability of the industry.

It is no coincidence that the concepts of nature-based tourism and biodiversity conservation evolved at roughly the same time. Ecotourism was a natural reaction to mainstream tourism which was perceived to be the major contributor to the degradation of the natural attractions on which it is based. Yet, from the point of view of the conservation of biodiversity, it doesn't matter at all whether visitors to an area are led by a trained or untrained guide, or whether the tour is classified as an ecotour or a large group day trip.

There is a chilling awareness spreading throughout the world of the urgent need for environmental protection and resuscitation. The spread of the conservation ethic, bolstered by the efforts of conservation groups, political initiatives, and the media, has influenced the choice of travel destinations. Tour operators are packing trips to rapidly disappearing wilderness and tourists are scampering to see areas which may not exist in a few years. Some have a fatalistic attitude and want a glimpse of the scenery before it fades away. Others recognise the link between tourism dollars and preservation, and want to support the conservation of these highly threatened areas.

A major motivation for tourism activities, both domestic and international, is to experience aspects of India's natural and cultural environment. Tourism development which exploits and degrades the environment is not only contrary to the principles of ecologically sustainable development (ESD), but is also likely to be ultimately self-defeating'.

In a positive light, the relationship between biodiversity and NBE can and should be mutually reinforcing. On the one hand, the declared and publicly promoted protection of natural features, ecosystems and biodiversity acts as a strong attractor for the tourism trade and provides a vehicle for the development of national and regional economies. On the other, there are opportunities—and indeed a strong obligation—for the tourism trade to promote and contribute to biodiversity conservation.

Tourism is recognised as a major user of biological resources while also providing employment for many people, supporting secondary industries, and contributing significantly to the economy. Benefits arising from the conservation of India's biodiversity are often grossly undervalued. The protection of natural areas assists in the provision of environmental services such as water supply, soil conservation, nutrient cycling and waste assimilation. There is an urgent need to apply positive methods of identifying and quantifying the benefits in order to demonstrate the value of natural ecosystems to government, industry and the community.

Numerous opportunities and benefits can be derived by strategically integrating biodiversity conservation requirements with future tourism needs. The goal of ecotourism is to capture a portion of the enormous global tourism market by attracting visitors to natural areas and using the revenues to fund local conservation and fuel economic development.

A large proportion of tourists attracted by the natural environment are in fact not seeking a 'total immersion' experience but rather wish to stay in an urban or resort setting and venture out for short visits to observe nature. While such experiences are not generally considered ecotourism, if effectively managed they can provide an efficient and sustainable means for relatively large numbers of people to enjoy some exposure to the natural environment with a much lower impact than would occur if they were to seek a 'total' nature-based experience involving camping overnight and other outdoor activities.

From the perspective of biodiversity conservation it makes little difference whether tourists come to be inspired by nature for an hour or a week. Furthermore, the same tourist who comes for a total immersion experience flies in the same aircraft, takes the same road, stays in the same accommodation and eats in the same restaurants as those that come to be inspired for an hour.

International Convention on Biological Diversity

The protection of biodiversity is an important national objective. India is a party to the International Convention on Biodiversity which was ratified on 18 June 1993 and came into force internationally on 29 December 1993. The overall objectives of the Convention on Biological Diversity are the conservation of biological diversity, the sustainable use of its components and the fair and equitable sharing of the benefits arising out of the utilisation of genetic resources.

The Convention has global scope. Each country which is Party to the Convention has responsibility for the conservation and sustainable use of its own biological diversity. Parties also have a responsibility to manage their own processes and activities which may threaten biological diversity, regardless of where their effects occur. Parties to the Convention are required to integrate conservation and sustainable use of biological diversity through the implementation of national strategies, plans and programmes for sectors such as agriculture, fisheries and forestry and for crosssectoral matters such as land use planning and decision making. Interestingly, tourism, a major user of natural resources, has not been specified as a sector.

Given that tourism in India is largely dependent on and a major user of natural resources and biodiversity, it is recommended that it be identified specifically as a sector in national policies that deal with biodiversity, conservation, ESD and the environment, where it is not already identified. Other important measures in the Convention are the enhancement of knowledge and understanding of biological diversity and the impacts on it.

Parties are required to identify and monitor important ecosystems, species and genetic components of biological diversity, as well as processes and activities that have or are likely to have significant adverse impacts on biological diversity. In this way, countries are able to determine their priorities with regard to conservation and sustainable use measures which need to be undertaken.

Parties are required to give emphasis to *in situ* conservation through a broad range of actions, including the establishment and management of protected areas; conservation and sustainable use of biological resources within and outside protected areas; promotion of environmentally sound and sustainable development in areas adjacent to protected areas; rehabilitation and restoration of degraded ecosystems; control of alien species and genetically modified organisms; protection of threatened species and populations; and regulation of damaging processes and activities.

Various measures are to be undertaken by Parties to promote sustainable use of biological diversity. These include integrating consideration of the conservation and sustainable use of biological resources into national decisionmaking; adopting measures for the use of biological resources which avoid or minimise adverse impacts on biological diversity; supporting local populations to develop and implement remedial action in degraded areas; and encouraging cooperation between governmental authorities and the private sector in developing methods for the sustainable use of biological resources.

The Convention provides for the adoption of economically and socially sound incentive measures, such as proper pricing of biological resources and the use of tradable rights in their management. Such measures can be an effective means of encouraging the conservation and sustainable use of biological diversity.

The Convention includes provision for encouraging public understanding of the significance of biological diversity and the measures for its conservation. Parties are to introduce appropriate procedures for environmental impact assessment of projects, programmes and policies that are likely to have significant adverse effects on biological diversity. The Convention also provides for the notification of activities which are likely to significantly damage biological diversity and the promotion of emergency response arrangements.

Protection of Natural Areas and Biodiversity

Biodiversity can be protected through a broad range of policies and specific measures. At the broadest level, ecosystems may be protected through the declaration of conservation reserves. Effective on-site management of reserves is essential to prevent or minimise potential adverse impacts of human or other activity. Codes of practice may be required for specific sectors, such as tourism.

Buffer zones and lands outside reserves are also used for conservation purposes. Land-use zoning, regulations, codes of management practice and other restrictions may all be used to reinforce conservation policies. Funding mechanisms and other commercial incentives may be introduced to encourage research and improve conservation practices. None of these approaches in isolation is likely to be fully successful, as each has its limitations. To achieve maximum success, it is important to adopt an integrated approach with careful coordination of a range of policies and protection measures.

Development of Planning Frameworks

Integrated regional planning, where environmental characteristics are a principal determinant of boundaries, is considered to be of major importance if biodiversity conservation is to succeed. Bioregional planning approaches are already being used in many regions. Planning at the regional level provides the appropriate scale to integrate reserve and off-reserve biodiversity conservation measures.

A key first step in the whole process is to identify a national system of bioregional planning units that emphasise regional environmental characteristics, based on environmental parameters such as vegetation types, catchment areas and climatic factors, and taking into account productive uses and the identity and needs of human communities.

While some countries are already applying bioregional planning approaches or its precursors, much work is still required to develop a national bioregional planning framework based on ecological, economic and social factors.

For example in Australia, work is underway to develop an Interim Biogeographic Regionalisation for Australia (IBRA). While the regionalisation is restricted to biophysical parameters and consequently has not considered economic or social factors integral to a national bioregional planning framework, it will be a key input in the development of the framework.

The IBRA has been developed specifically as a framework for setting priorities in the National Reserves System Cooperative Programme. It comprises 80 terrestrial biophysical regions and is the result of on-going collaborative efforts of Commonwealth, State and Territory nature conservation agencies. The regions are defined on a landscape-based approach to classification of the land surface, including attributes of climate, geomorphology, landform, lithology, and characteristic flora and fauna. IBRA was derived from the best available continental scale data, published reports, and biogeographic regionalisations for each State and Territory.

In terms of biodiversity the IBRA regions are at best a convenient approximation of the complexity observed in the real world. The assumption that IBRA is a predictor of overall biotic regionalisation and regional biodiversity has not been tested. Faunal distributional data, for instance, have made a relatively minor contribution.

IBRA is appropriate for use only at the continental landscape level. More detailed information is required to identify regional and local scale patterns, keeping in mind that the assessment of heterogeneity within regions will be biased by the level of information available. Well studied regions will appear more heterogeneous. Of direct relevance to tourism is the fact that IBRA does not take into account special values which may include outstanding natural features, cultural values and landscape values.

One of the best examples in Australia of planning to integrate biodiversity conservation with various uses at the bioregional level, especially tourism uses, is the Great Barrier Reef Marine Park. Since the beginning of its declaration in 1975, and subsequent reservations, the park has undergone intensive planning by the Great Barrier Reef Marine Park Authority (GBRMPA), in conjunction with all users and management authorities, including the Queensland National Parks Service Several successional plans have been developed, resulting in the delineation of 4 sections and management zones for conservation, recreation and commercial utilisation.

The compromises involved in determining these zones have resulted in better understanding of management issues by all users, and a satisfactory approach to the conservation of this vast ecosystem. In preparation of a strategic plan for the Great Barrier Reef World Heritage Area, more than 60 major interest groups were consulted and a 25 year vision for the area, and 5 year objectives were agreed to by the groups. This could be the first time such a large area involving multiple marine and terrestrial jurisdictions has been subjected to such an effective coordinated approach.

One of the reasons this has been possible is that the whole of the reef, with some minor exclusions, is under one management authority, enabling competing uses, such as conservation, tourism and fishing to be satisfied while not compromising the high conservation values of the reef.

National Parks and Biodiversity

Park and reserve declaration has been largely ad hoc and based to a large extent on amenity or landscape and recreational values, rather than on intrinsic biodiversity values or identification of deficiencies in reservation of representative samples of ecosystems. Biodiversity values have received higher priority over the past decade or so, as show by significant reservations such as the Kaziranga National Park, Keoladeo Ghana National Park, Manas Wildlife Sanctuary, Nanda Devi National Park, and Sundarban National Park

The parks and reserves estate includes areas owned and/or managed by the state. While the importance of the national parks and reserves should not be understated, some of these parks may be of significance only at the local or regional scale, and not for their representativeness of biodiversity values.

At present, the system of national parks and protected areas in India is the principal means of conserving biodiversity. Yet reserved areas do not adequately conserve India's biological diversity, and a number of identified ecological communities are in critical need of protection. It should be recognised that conservation of biodiversity will always depend on conservation both inside and outside reserves, and that land owners and local people need to be involved in this conservation. Identification and declaration of particular national parks and natural areas often behaves as a magnet for tourists, who are attracted by the very selection of those areas and the fact that the areas are made open to public access.

Currently, NBE is almost totally dependent on India's protected area system. In other parts of the world as well, NBE is dependent on protected area systems. The World Resources Institute, in recognising the importance of tourism to the protection of areas, notes that 'although some 7,000 protected areas exist throughout the world, comparatively few enjoy *de facto* protection, and most of those that do can attribute their survival to the revenue they earn from tourism'.

Local communities may have mixed feelings about declaration of 'their' lands to national parks. Attitudes are often negative initially due to the assumption that the land will be 'locked up' from mining, hunting, grazing and logging, but later develop into a positive one through being associated with an important national park, which also generates significant regional economic benefits. Attitudes of the public and elected representatives can change to become supportive, at least in part or in principle. It is recommended that creation of a representative system of protected areas be accelerated, being crucial to the protection of biodiversity. The process of establishing such a system is an essential component of ecologically sustainable development.

The maintenance of biodiversity outside protected areas has received too little attention. In the face of significant and continuing reductions to our biological diversity, there is a pressing need to strengthen conservation activities across India. About 70 per cent of India's land area is under the control of private landholders and resource managers, including indigenous peoples; their cooperation is essential for the success of conservation activities. High priority must be placed on developing and implementing integrated approaches to conservation that both conserve biological diversity and meet other community objectives.

Although many programmes exist to encourage better management of biological diversity on such lands, greater effort is required to raise both the standards of management and protection and the levels of financial and technical assistance'. Actions which could help achieve the objective of off-reserve conservation of biodiversity involve financial incentives to property owners affected where areas of significance to biodiversity are protected. These include the negotiation of conservation covenants and heritage agreements between owners and managers and governments, which in turn provide resources.

From a business perspective, a system of secure areas that will still be there after several decades encourages investment in land and promotion for tourism. In the past this has usually meant that investment follows only after the dedication of areas as national parks, since this offers a guarantee that the resource will be there indefinitely. Private landowners, however, who are prepared to enter into legally binding conservation management agreements with government authorities could also benefit from investment opportunities. This approach has the added benefit of making funds, available for the acquisition of biologically important parcels of lands, go further.Information packages detailing requirements for management for conservation objectives could be developed for distribution to landholders. These packages would explain the NBE/off-reserve conservation option which would also include assistance in developing an NBE business, such as training, product development and marketing.

It is recommended that a national system of NBE/conservation covenants be investigated which involve legally binding Conservation Management Agreements for private lands with high conservation values and sympathetic managers/owners who wish to develop a tourism enterprise. Part of the revenue from such enterprises would go toward managing the private reserve. Lands covered by such agreements should be given support for:

— the rehabilitation of lands, for example in the establishment of native vegetation corridors;
— funding for necessary restoration programmes;
— technical support and the provision of appropriate seed stocks;
— development of a monitoring and reporting programme to determine the effectiveness of rehabilitation; and
— the integration of data into national park and national monitoring databases.

In addition, assistance should be given for the development of NBE enterprises. Such assistance would include business planning, training, product development and marketing.

Use of Information Systems for Biodiversity Conservation

Biodiversity conservation is carried out largely through the establishment and management of the system of protected areas. In the past the reservation of protected lands has been largely *ad hoc* and opportunistic, resulting in a reservation system which is not representative of biodiversity.

The primary challenge for biodiversity conservation, therefore, is the development of a representative reserve system based on regional biodiversity. While few, if any, comprehensive regional ecological databases exist, systematic approaches for determining representative systems for regional biodiversity conservation are rapidly being developed. In most regions there are many possible ways of combining numbers of sites into reserve systems to accommodate changing land uses.

Environmental Impacts of NBE

The range of impacts and potential impacts of tourism includes such examples as direct damage to coral reefs, overfishing of estuaries and lakes, damage to shorelines from boats, water and soil pollution, disturbance to turtle and bird nesting sites, over-collecting of firewood (sometimes by park managers), spread of toilet paper in bush camp situations where there are no constructed toilets (the paper should be burned), water pollution from campers, faecal contamination of sites, litter accumulation, track erosion and damage, plant collection and flower picking, disturbance of animals in their habitats, trenching around tents, cutting of vegetation for shelter construction, littering, overhot fires, vehicle damage off formed or formalised tracks, touching of prehistoric art, collecting souvenirs of natural objects and artefacts, sign damage, barrier damage, infrastructure development including roads, power and water reticulation, buildings, and rubbish pits, and, of course, the management practices applied to correct the impacts.

Properly planned and managed tourism is as important to minimise impact and lead to 'positive environmental benefits'. There are inadequacies in current tourism management practices, which have led to a depletion of natural resources and destruction of important representative habitats. Some of the practices needed to reduce impacts of ecotourism, including codes of conduct, education on ecological sustainability, management of operators by licensing and accreditation, research and monitoring of impacts, and a number of others.

Many of the environmental problems result from a lack of awareness and understanding in the tourism industry about how people affect the environment and what they can do to lessen their impact. Well-planned information, education and marketing strategies can help increase knowledge and understanding and also influence environmental decisions, attitudes and behaviours.

The concepts inherent in ecotourism hold the promise of introducing a new ethic in tourism, one that is sensitive, responsible and conservation-minded. While this ethic is evident in some NBE enterprises, it certainly does not apply to all within this segment of the industry. Nature tourism is not necessarily ecologically sound just because it uses nature as its primary resource, as numerous examples around the globe illustrate. NBE has frequently simply been a marketing tool for many businesses, and the negative impacts of tourism are as evident in nature-based and ecotourism as in any other.

Management of Environmental Effects

Management of biotic and abiotic resources is implicit in all aspects of an ecologically sustainable economy, including tourism. Management of the resources is not a simple concept, nor an easily achievable practice, but requires the development of broad national strategies and the implementation of those strategies, national and regional level.

Governments and participants in the tourism and recreation industry can help conserve biological diversity by:

- Reviewing the impact of tourism and recreation management activities on biological diversity and seeking changes where appropriate;
- Developing and implementing for tourism operators using areas with significant biological diversity, codes of practice that acknowledge the need for any required changes to management practices; and
- Offering incentives for conservation activities, including rehabilitation programmes.

Where tourism is dependent on the natural environment, management strategies should be developed in association with broader land use plans, including provisions for:

- Tourism facilities and services to be provided in accordance with the biophysical limits of an area;
- The development of criteria and conditions under which commercial activities within or adjacent to protected areas may be appropriate; and
- Rehabilitation of existing tourism sites where appropriate.

Establishing a representative system of reserves is essential to the industry and to conservation of biodiversity, but that of itself is not sufficient. The system of reserves is inadequate to conserve total biodiversity, and will always be so. Management of biodiversity must be implemented across the broader landscape.

Ecotourism and nature-based tourism occur across the world, and, by inference, utilise the natural landscape and its component species and habitats. By extension, then, the management of the impacts of tourism on biodiversity conservation must extend beyond the boundaries of parks and reserves. In light of the numerous strategies relating to ecological sustainability and biodiversity conservation of recent years, this is a surprising and unfortunate omission and should be rectified for future strategy development and implementation.

The tourism industry has a very proactive role, a duty and an economic incentive to communicate ecologically sustainable practices and management needs to the huge numbers of tourists they convey and to whom they interpret. After all, tour guides have a captive audience and an opportunity to teach more people about the environment than all universities and most schools. Their role in conservation of biodiversity has been under-rated for a long time.

The industry as an entity does not yet consider that management of the natural resource should be part of its responsibility. Apart from a few easily identified operators, most consider that management is the sole responsibility of the 'management authority'. A caveat to this is that there are a good and increasing number of park management boards and advisory committees in which members of the tourism industry participate, but they consider themselves and are considered as advisers on tourism and conduits for information, rather than as active participants in management. By active participation we do not mean necessarily physical management works; but we do mean an active and legitimate role in planning and determining management practices.

For the goals to be translated into action, the industry must become involved in this process: through management boards, through direct consultation with managers, by learning about impacts, and through notification to relevant authorities of identified management problems. There will never be sufficient funds nor resources for management authorities to manage all impacts adequately without the involvement of the users of the resources.

For this to happen, a shift of attitude in the authorities is also required. Too often, managers consider themselves the prime authorities on park management. While we acknowledge that most are skilled and highly trained in management, there are cases where the tourism operators also see and can identify problems on site, sometimes problems which managers are not able to see, for instance, because people are more careful around park managers.

Some environmental problems which operators can identify might include slow site deterioration which short-term managers may not identify, changes in the quality of a site, such as change in bird behaviour, change in health of vegetation, overcrowding, physical damage to or subtle physical impacts on prehistoric art and cave formations, flower picking, inappropriate practices such as use of detergents or soaps or even washing in streams and waterholes, and myriad others. This shift is progressively happening facilitated by the increasing levels of consultation with a wide cross section of the community which now occurs in protected area planning and management.

Some means of overcoming these management needs include workshops and training courses which are already conducted by park management authorities, and sometimes by industry bodies. While recognising the value of these courses and workshops, they are often designed and conducted by the authorities and associations as 'training' and 'orientation' courses for operators and guides, rather than as means of facilitating two-way interaction, and mutual resolution of management and operational problems. There is a need for workshops to be conducted which address management of the resources for the benefit of both the natural environment and the industry. The workshops should be short and frequent, and focus on key issues, such as means of identifying management problems, and developing solutions to problems.

Strategic Planning For NBE Development

Identification of Areas for Detailed Planning

If tourism is to develop in an ecologically sustainable way, the current political and institutional fragmentation existing in landuse planning will need to be overcome. Regional planning should focus on integrated landuse plans based on ecological systems or biophysical regions. In the context of regional plans, strategic tourism plans should be developed by government in collaboration with local governments and tourism industry bodies which would, amongst other things, identify ecologically appropriate areas for tourism use and development.

Assessing Regional Economic Impacts of NBE

There are well established methods for assessing the economic implications of NBE tourism development and for conducting development planning. The usual framework is an input-output or multi-sectoral economic model, designed to be applied at regional state or national level.

Multi-sectoral models vary in sophistication from fairly simple representations of the economic structure to large-scale simulation models with many policy and other variables and complex mathematical functions connecting them. It is important to note that the economic variables incorporated in the models are based only on market transactions and do not include broader community economic values such as existence, bequest, option and quasi-option values.

Tourism is not an identifiable sector in the standard classification of industries. Rather, it places demands on a wide range of industries which include food, beverages, accommodation, retail trade, transport, entertainment and personal services. Part of tourism expenditure consists of taxes and charges paid to governments such as airport taxes, park use fees for national parks, bed taxes and user charges paid to local councils for local services such as roads, waste disposal, water and sewerage.

Nature-based and ecotourism requires quite detailed expenditure data, as they comprise a special major component of the total tourism market. At present, no data set representing specific patterns of nature-based and ecotourism expenditure has been compiled. It should be a high priority in strategic planning for ecotourism development.

One of the important applications of multi-sectoral models in evaluating the role of ecotourism is to make a baseline assessment of the current dependence of the economy—regional, state or national—on the ecotourism trade and of specific industries within the economy in terms of their total turnover, income and employment.

A second application is to predict the impacts of changes in the level and structure of the tourism trade, based on scenarios or forecasts of future trends in tourism markets.

A third application is to use the results of model simulations to formulate regional development strategies, with a focus on, among other things:

— Opportunities—both direct and indirect—for industry to benefit from a future expansion of tourism;
— Requirements for public and private investments in infrastructure;
— Demands for and availability of labour in specific skill categories;
— Fiscal implications for local and state government;

— Facilitation and coordination of industrial development associated with the tourism trade.

Applying Multi-sectoral Models

Extensive work has been undertaken on conventional regional economic models, but new directions are needed in integrating this kind of planning with bioregional planning. To apply multi-sectoral modelling in strategic planning for ecotourism development the spatial location and boundaries of key bioregions must be matched with corresponding regional economic data.

Market studies can be applied to determine likely growth rates for tourism in each region and the demands for specific kinds of goods, services and infrastructure associated with the tourist trade. Impacts on regional economies can then be predicted using standard modelling techniques. The results of model simulations can be used to assess the feasibility of development plans and environmental protection programmes for each key bioregion and linked regional economy.

The economic model is constructed from base year data indicating the flows of expenditure between sectors of the economy; sales to consumers, government and export markets; and expenditures on primary inputs to production such as labour and capital. The data must support the mathematical form of the model.

The simpler models are based on the assumption that consumer or other demands are pre-determined, that all inputs to each sector are required simultaneously in fixed proportions and that substitution of inputs does not occur. Such models predict the pattern of industrial activity that is required to meet the prescribed set of final demands. The more complicated models allow for input substitution, changes in the pattern of demand and other more complex economic effects.

The next step is to assess the economic implications of nature-based and ecotourism. It is necessary to identify the pattern of expenditure for this segment of the tourism market. Data are required for key components of expenditure such as guided tours, transport, food, accommodation, photographic supplies, equipment and personal services, each associated with a particular economic sector. Once these data have been collected it is possible to assess the dependence of the economy, specific industries, and the work force on the ecotourism market.

Although the methodology is straightforward, there are currently severe limitations on the availability of data, particularly on expenditure patterns associated with nature-based and ecotourism. Indeed, data on expenditure patterns for tourism in general are currently extremely limited. The expenditure components are usually limited to commercial accommodation and daily expenditure. More is known about the expenditure patterns of international visitors. To conduct reliable assessments of the economic importance of nature-based and ecotourism, extensive primary data collection on expenditure patterns is needed. Such data should cover not only actual expenditure currently made by ecotourists, but also the kinds of natural areas, services or activities tourists would like to spend their money on in the future.

Planning for Infrastructure Support

The infrastructure needs of the ecotourist differ from the mass market tourist primarily in requirements for the locality and style of accommodation, transport services, and access to sites. Ecotourists seek local vernacular accommodation that allows them to be close to the natural environment rather than removed from it. They seek designs and styles which expose them to the local culture, not cosmopolitan internationalised architecture.

Infrastructure needs include booking systems and infrastructure outside National Parks. Even where infrastructure developments are undertaken largely by the private sector, Governments can assist by identifying suitable sites, assisting developers with concept development, entering joint ventures such as information centres, and assisting with funding where public land is controlled by a local committee of management.

People have broad tastes and preferences in accommodation and other infrastructure, and will avail themselves of their preferred style if it is available. In planning for infrastructure support, planners must consider the place where the developments are proposed, the range of infrastructure currently available, the ecological/environmental constraints and opportunities, the purpose and goals of the development, and the potential market and proposed activities for the area. The infrastructure must be appropriate to the area.

In formulating strategic development plans, it is important to distinguish the economic effects of expenditure directly attributable to tourism, which will tend to be ongoing, from effects that result from capital expenditure in supporting infrastructure, which may be only of a transitory nature. The various spheres of infrastructure investment should be identified, such as local, state and central government and the private sector. Care should be taken to avoid the 'boom-town' effect whereby regional production, income and employment increase rapidly during the capital construction phase, but decline when required facilities and infrastructure have been provided. The provision of information on predicted patterns of development should assist industry to plan new development projects and help government agencies to plan for public investments in supporting infrastructure.

References

Batta R.N.(2000). *Tourism And The Environment: A Quest for Sustainability*. Indus Publishing Company, New Delhi. pp. 203.

Burger, J. (2000). "Landscapes, tourism, and conservation". *Science of the Total Environment*. 249 (1-3): 39-49.

Cater, E. (1994). Cater, E., and G. Lowman. ed. *Ecotourism in the Third World — Problems and Prospects for Sustainability in: Ecotourism, a sustainable option?*. United Kingdom: John Wiley and Sons.

Ceballos-Lascurain, H. (1996). *Tourism, Ecotourism and Protected Areas*. Gland, Switzerland: IUCN.

Charters T. (1996). Ecotourism: A tool for conservation. eds. Chatters T. et. al. *National parks: Private sector's role*. Toowoomba, Queensland, Australia: USQ Press. pp. 77-92.

7

Ecotourism in Protected Areas

By definition, ecotourism is about traveling to and visiting natural areas, places where nature still exists in a relatively unaltered state. In a world where population pressure and increased resource consumption are placing huge demands upon our natural resource base, natural areas are increasingly hard to find. At the same time, our global cultural heritage is under attack, making it increasingly difficult to learn from other cultures and to remain in touch with cultural roots throughout the world. Today, the remaining natural areas are mostly protected in some way. Ecotourism attractions, whether they are wildlife viewing possibilities or dramatic natural landscapes, tend to be found in these protected natural areas.

Protected areas began evolving in the 19th century largely as a response to these pressures. By "protected area" we mean a piece of land (or body of water) which is characterized by the following:

1. The area has defined borders.
2. The area is managed and protected by an identifiable entity or individual, usually a government agency. Increasingly, though, governments are delegating responsibility for protected areas to other entities that are private, public or a combination thereof.
3. The area has established conservation objectives that its management pursues.

The rapid increase in the numbers and territorial coverage of protected areas since the 1960s coincides with more rapid increases in the aforementioned pressures. Traditionally, protected areas are set aside and managed by government authorities in order to protect endangered species or examples of outstanding scenic beauty. In much of the southern hemisphere, financial pressures on government budgets, global trends towards decentralization and a society which increasingly values the role of nongovernmental participation have caused some profound changes in the way protected areas are being administered and managed.

These changes are manifested in two major ways:

1. Protected areas are increasingly expected to generate some portion of the funding necessary for their own management.
2. Many other organizations, both private and public, are becoming involved in the management and conservation of protected areas, either in partnership with the traditional government agencies in charge of protected areas or by managing their own protected areas.

An additional responsibility of park managers is to bring conservation to the people. Without a constituency for conservation, we will ultimately fail. This constituency can be local, national and international. Ecotourism is crucial for achieving this goal and not just as a source of conservation finance. The link between ecotourism and protected areas is therefore inevitable and profound.

The Role of Ecotourism

Tourism and ecotourism are usually a part of the management strategy for a protected area. The degree to which tourism activities are pursued depends upon the priority assigned to them by the area managers, who in turn should be guided by a planning document prepared for that purpose. The planning document (or management plan) should be the result of a comprehensive evaluation of the area's natural and cultural resource base. It determines the stresses, their sources and the real threats to the area's natural and cultural integrity, as well as the strategies to reduce these threats. The plan should define the area's long-term management objectives and a zoning scheme that identifies where certain activities may take place.

What we have is a coming together of two different forces to create a symbiotic relationship: ecotourism needs protected areas, and protected areas need ecotourism.

Ecotourism is increasingly being considered as a management strategy for protected areas that, if implemented appropriately, constitutes an ideal sustainable activity. It is designed to:

— have minimum impact upon the ecosystem;

— contribute economically to local communities;

— be respectful of local cultures;

— be developed using participatory processes which involve all stakeholders; and

— be monitored in order to detect negative and positive impacts.

There are many compelling reasons why conservationists and protected area managers are considering ecotourism as a protected area management tool. These include the following:

1. Conventional tourism sometimes appears as a source of stress on the biodiversity of a protected area. In other cases, ecotourism can be regarded as an appropriate strategy for addressing threats to conservation targets. Nature tourists are presently going to protected areas in growing numbers. At a minimum, managers must control tourism's negative impacts. Even if elaborate visitor centers and extensive tourism businesses are not created, measures must be taken to ensure that these growing numbers of visitors do not negatively impact the biodiversity values of a protected area. These measures include increasing staff, developing monitoring systems and refining environmental education efforts. Managing visitors and minimizing impacts is a primary responsibility of protected area managers.
2. Ecotourism can capture economic benefits for protected areas. Visitors with no place to spend money are missed opportunities. Hundreds of thousands of dollars of potential revenue currently are being lost both to protected area managers and local communities because tourists do not have adequate opportunities to pay fees and buy goods and services.
3. Properly implemented, ecotourism can become an important force for improving relations between local communities and protected area administrations. This relationship is perhaps the most difficult aspect of ecotourism since it involves levels of communication and trust between different cultures and perspectives that have traditionally been difficult to achieve.

4. Ecotourism can provide a better option than other competing economic activities for natural areas. Many natural areas are threatened and need to be fortified in order to survive; ecotourism may help guard against some of these threats and competing land uses. For example, a successful ecotourism program can forestall implementation of logging in an area by generating greater revenues, especially over the long term.
5. By implementing ecotourism in protected areas, we are demonstrating that tourism need not be massive and destructive. We are demonstrating that, even within the fragile environment of protected areas, sustainable development can work.

Opportunities and Threats

Tourism presents a mix of opportunities and threats for protected areas. Ecotourism seeks to increase opportunities and to reduce threats. If an opportunity is realized, then it becomes a benefit. If a threat is not avoided, then it becomes a cost. There are no automatic benefits associated with ecotourism; success depends on good planning and management. Carelessly planned or poorly implemented ecotourism projects can easily become conventional tourism projects with all of the associated negative impacts.

Opportunities and threats, and consequently benefits and costs, will vary from situation to situation, from group to group and from individual to individual within groups. Benefits to one group may be costs to another. Determining which opportunities to pursue and which threats to abate is a subjective decision that can best be made by involving all stakeholders. Ranking the importance of each benefit is part of the compromising involved in the ecotourism planning process.

The entire spectrum of ecotourism's opportunities and threats does not apply to every protected area. For example, in a protected area that attracts primarily domestic visitors, opportunities to generate foreign exchange are limited, but good opportunities may exist to raise conservation awareness locally. Environmental degradation will vary depending on the fragility of natural resources and the types of activities that are permitted. The circumstances of each protected area create a particular set of opportunities and threats.

The remainder of this chapter identifies and describes the opportunities and threats that tourism development represents for a protected area.

Potential Opportunities of Ecotourism

Bringing money into protected areas is a major concern of conservationists. Governmental funds available for protected areas have been decreasing globally, and many important natural areas will not survive without new sources of revenue. Tourism offers opportunities to generate revenue in diverse ways, such as entrance fees, user fees, concessions to the private sector and donations. New funds allow protected area managers to handle tourists better and to hold the line against other threats.

Entrance or visitor use fees are charged directly to visitors to see and experience an area. Collected at the gate, entrance fees have various structures. In some cases, a flat fee is charged. In other cases, multi-fee systems are established with various rates for different types of users. Typically, foreign tourists are charged more than local visitors are. User fees are charged for specific activities or for using special equipment in a protected area, such as electrical hook-ups when camping or various rental fees.

Private sector concessions include snack bars, restaurants, lodges, gift shops, canoe rentals and tour guides. All of these can be privately owned or managed with a portion of the profits returned to the protected area. This arrangement is favorable because it reduces business responsibilities assigned to untrained or uninterested protected area personnel. Concessions allow protected areas to benefit from the energy and profits of private sector enterprises. However, concessions must be negotiated for the protected area's long-term benefit and must be monitored closely. This monitoring ensures, for example, that the concessionaire is complying with contracted services such as trash removal, trail maintenance, etc.

Donations may be solicited through a simple box at the door or perhaps via a more sophisticated campaign such as an "adopt-an-endangered-species" program. Protected areas with threatened or unique plants and animals can request financial assistance for them. Visitors who have just completed a fascinating nature experience are a perfect audience for this type of appeal. Many protected areas report a high rate of success with setting up donation programs for specific campaigns. For example, Fundación Natura in Colombia and ANCON in Panama have successful "adopt-a-hectare" programs. The Galapagos Islands National Park has a successful "Friends of Galapagos" program. Such programs and funds should be established parts of any ecotourism program to a protected area. Ecotourists want to contribute to conservation — let's not deny them the opportunity!

There may be other ways tourism can bring revenue to protected areas. For example, visitors may also be "virtual," which entails visiting a web site that has been established for a protected area. Donations may also be solicited from a much larger audience of such virtual visitors. For some protected areas, tourism can become the primary revenue generator. For others, it will be only one of many sources of financial contributions. But for almost all protected areas, visitors should be considered a readily available and accessible income source that should be exploited equitably for long-term sus-tainability and to promote return visits.

Employment Creation

New jobs are often cited as the biggest gain from tourism. Protected areas may hire new guides, guards, researchers or managers to meet increased ecotourism demands. In surrounding communities, residents may become employed as taxi drivers, tour guides, lodge owners or handicraft makers, or they may participate in other tourism enterprises.

In addition, other types of employment may be augmented indirectly through tourism. More bricklayers may be needed for construction. More vegetables may be needed at new restaurants. More cloth may be needed to make souvenirs. Many employment sources are enhanced as tourism grows.

In some cases, community residents are good candidates for tourism jobs because they know the local environment well. Residents are ideal sources of information; for example, they can tell visitors why certain plants flower at particular times and what animals are attracted to them. As indigenous residents of the area, community members have much to offer in ecotourism jobs. However, care must be taken to protect the rights (sometimes referred to as intellectual property rights) of local peoples so that their knowledge is not exploited or appropriated unfairly by visitors or a tourism program.

The Kimana Group Ranch, outside of Amboseli National Park in Kenya, gained international attention for establishing the first community wildlife sanctuary in Africa. Managed by the Masai ranchers, Kimana has its own warden, guides, entry gate and lodge concessions.

We should not overstate the value of ecotourism employment in rural areas. There are a few important caveats to consider. First, while there is often talk of big tourism dollars, ecotourism will generally not be an economic bonanza for an entire community. More realistically, it will generate some jobs, depending on how popular the protected area is, but will not automatically become an income provider for hundreds of people. Furthermore, many ecotourism jobs will be part time and seasonal and should be considered only supplemental to other sources of income. Overall, ecotourism employment will likely be limited for most communities.

A second concern about ecotourism employment is the nature of jobs for communities. Typically, few management and ownership positions are available. Tourism will always have many service positions, because it is a labor-intensive industry. But communities may resent ecotourism if their members are not represented in the higher levels of employment. The profitability of tourism for local residents is minimized if they are offered only menial jobs and not given opportunities for advancement. Additionally, gender inequities may be generated while the higher paying guiding and management jobs all go to men and women are restricted to lower paying laundry, cleaning and cooking positions.

Another hurdle to ecotourism employment is the issue of training. For many residents, new employment is a major personal and professional transition. It sounds good on paper that former loggers may become tour guides, thereby conserving the trees they used to cut. But redirecting careers is a big undertaking. New job candidates need information on all facets of ecotourism management. They need training in business development as well as such basics as languages, food preparation, first aid, motorboat maintenance, interpretation, group management, etc. They need access to international markets. New tourism jobs require new skills and therefore training. Ecotourism project plans need to budget for these training costs over the long term.

In addition, there are many social and cultural considerations in switching jobs; it involves lifestyle changes.

Diversifying into nature tourism jobs may change the way communities look and operate. Conflicts among residents may develop. For example, tourism jobs are likely higher paying than traditional sources of income. Within a community, a farmer may earn the equivalent of US$50 a month. A neighbor working as a tour guide may earn the same amount with one tip from a wealthy tourist. Will these inequalities create jealousies? How are these resolved? Who gets the coveted tourism jobs if there are more candidates than opportunities? Does a community want to become a tourist destination if it means losing traditional economic foundations, such as agriculture?

One important issue to keep in mind when evaluating the effectiveness of ecotourism jobs is what employment alternatives the local populations have. In many cases, ecotourism may be the best option if the other potential land uses are more threatening to the survival of the area's natural resources, even if these ecotourism jobs are few and flawed. In analyzing ecotourism jobs, it is essential to keep in mind their relationship to threats to the biodiversity of an area.

Justification for Protected Areas

Visitors, or the potential to attract visitors, are among the reasons that government officials and residents support protected areas. For government officials, declaring areas protected and providing the financial assistance to maintain them is often a difficult process. These officials face many competing interests in making decisions about how to use land and marine resources. Conserving protected areas requires long-term vision; this is often a challenge for government officials, especially when confronted with the prospect of short-term financial gains for logging, mining and agriculture activities.

But as government officials review land and water-use options, nature tourism may sway them to provide protected status to an area or strengthen the protective status of an existing protected area or reserve, particularly if it can generate income and provide other national benefits. International tourism motivates government officials to think more about the importance of managing natural areas. Visitors are more likely to visit and support a natural area if it is protected, which in turn adds justification to the existence of protected areas.

Visitation to the area may be the impetus for residents near protected areas, or potential protected areas, to support the continued protection of these areas.

A Stronger Economy

Tourists visiting nature sites boost economies at the local, regional and national levels. If tourism brings jobs to residents at the local level, they then have more money to spend locally, and economic activity within the area increases. The same pattern may occur at the regional and national levels. Nature tourists arrive in the capital city of a country. They may stay for a few days or travel to the countryside. Along the way they use hotels, restaurants, shops, guide services and transportation systems. Typically, a multitude of businesses benefit directly from nature tourists. Although these businesses usually are set up to accommodate the broader groups of international and national tourists, nature tourists are an added market. Also, some operations whisk visitors directly from the airport to a full itinerary in a private protected area, thus leaving the visitor no opportunity to spend money in local communities. In such cases, it is important to ensure that there are mechanisms such as airport taxes to obtain at least some tourist revenue. Industries that support tourism, such as manufacturing and farming, are also affected by numbers of tourists. Growing ecotourism creates a stronger economy throughout the country.

National governments can also generate tourism dollars through import duties and taxes. For example, researchers determined that the Belizean government earned BZ$7 million from taxes on fuel used in the tourism industry (Lindberg and Enriquez, 1994). Other taxes include occupancy taxes (directly to hotels) or departure taxes (directly to tourists). These taxes are generally a good way to target visitors directly while avoiding inflationary problems with local populations. Also, these charges need not adversely affect demand. For example, nature tourists do not stay away from Belize because they have to pay a US$22.50 departure tax. This income is a big help to the national economy, with portions supporting the protected area system.

Environmental Education

Nature tourists provide an ideal audience for environmental education. During an exciting nature hike, visitors are eager to learn about the local habitats. They want to hear about animal behavior and plant uses as well as the challenges of conserving these resources. Many want to know the economic, political and social issues that surround conservation.

Nature guides are one critical source of environmental education. Visitor surveys show that good guides are a key factor in a trip's success. For example, in 1996 the RARE Center for Tropical Conservation asked 60 conservation groups in Latin America to identify their most urgent obstacle to developing ecotourism; the lack of well-trained nature guides ranked second in their concerns.

Visitor centers with displays, printed materials and videos are also an excellent means of environmental education. Additionally, interpretation in the form of trail signage can give important biological information and conservation messages. Interpretation for visitors is becoming increasingly creative and interactive.

Environmental education is an equally important opportunity to reach national visitors. Whether they are local school children learning about the resources that are valuable in their daily lives, or travelers from neighboring regions learning about the significance of their national protected areas, citizens are a key audience. Conservation messages have a special urgency for them.

Environmental education is most effective when pre and post-trip information is made available. Preparation encourages visitors to think about appropriate behavior, thereby minimizing negative impacts, and the use of follow-up materials continues the environmental education process.

Appreciation and Pride

Appreciation and pride are less tangible benefits than the others listed here, but they can lead to tangible actions. It is common for people not to fully appreciate their surroundings and to take what they have for granted. Often, it is outsiders who take a fresh look and add value to our resources. This phenomenon happens both in big cities and in remote natural areas. Although rural residents who have grown up among spectacular wilderness areas generally understand the intricacies of nature and value its role in their lives, many have little idea of the global importance of their natural resources. Many rural people do not realize the magnitude of the global attention, study and concern that their homelands receive.

On the other hand, adventurous nature tourists are often wildly enthusiastic about exploring new wilderness sites. They pour into small communities with video cameras and document all they see. Journalists from National Geographic and other magazines write inspiring stories with glossy photos. Natural sites that were once secret, especially in tropical countries, are being promoted with unprecedented fervor.

Native peoples are often surprised at the level of outside interest in their natural resources and in their culture. In most cases, however, they see their surroundings in a new light after international exposure. They gain a new appreciation for the nearby natural areas and wildlife that attract tourists. If the tourism experience is managed with proper community participation and control, it can also lead to greater appreciation by a community of its own culture, the same culture which visitors increasingly seek to learn about and admire.

Improved Conservation Efforts

As a result of growing appreciation and pride, conservation efforts often increase. Many residents are motivated to protect their areas and may change their patterns of resource use. Cultivation practices may be altered. Litter on roads may be cleaned up. Water may be better managed. Local populations often learn more about conservation and modify their daily habits because of tourism.

Awareness often increases at the national level also, resulting in such improved conservation efforts as mandating and supporting protected areas. Even at the international level, ecotourism may engender an international constituency for improved conservation efforts and support for particular protected areas. International and local visitors to a protected area are likely to rally to its defense if a valuable area is being threatened. For example, when illegal oil exploration was taking place in the Cuyabeno Wildlife Reserve in Ecuador in 1993, indigenous Quichua and Cofan communities which were very involved in ecotourism turned to environmentalists and tour operators in the region for support. The tour operators encouraged their guests to participate in what became a decisive campaign of international letter writing to stop the threat to the reserve and to the livelihoods of the local communities.

Issues of Tourism in Protected Areas

Protected areas were first set up during the 19th century. They were funded by governments and maintained as assets for their nations and populations. When first established, populations were smaller than today's, transport links to protected areas were generally poor, and general population pressures on protected areas were far lower than at the present time. As populations have grown and pressures on the environment and wildlife have increased, so too the importance of protected areas, and of conserving biodiversity, has been increasingly recognised. At international level, this has led to the development and implementation of the Convention on Biological Diversity, and to an increase in the number of nationally-designated protected areas to around 100,000. Over the same period, pressures on protected areas have increased dramatically, and include pressures to open these areas for agriculture, forestry and oil and mineral extraction. If the 100,000 nationally-designated protected areas each require an average of USD 500,000 per year for their maintenance, the total cost of maintaining the global network of protected areas will be around USD 50 billion per year and the cost of conservation of 25 of the top world biodiversity hotspots has been estimated at USD 500 million a year· However, very few protected areas generate sufficient income to cover their maintenance costs at present.

Protected areas have traditionally relied on support from the government, but the costs of conservation while relatively low, represent a significant amount for governments, particularly those in developing countries and those with economies in transition, to provide. Revenue from government sources is now in short supply, and the need to balance environmental and economic requirements is becoming increasingly pressing. There is an urgent need to find ways for protected areas to be able to generate sufficient funds to make a contribution to their operating costs. The CBD's Programme of Work on Protected Areas highlights the need to foster economic opportunities based on protected areas, and for development of mechanisms to help protected areas achieve financial sustainability. There are some examples of protected areas that are successful in raising substantial amounts of revenues from various non-government sources, through for example, foundations, and entrance and user fees: however, these are a minority of protected areas.

In today's world, protected areas are increasingly expected to 'pay their way', either by generating their own revenue, or by stimulating sufficient revenues at regional or national levels for example, through the tax system to pay for a significant proportion of their maintenance costs. In addition, where protected areas are under threat as a result of pressures to convert them to alternative resource uses, their survival can depend on demonstrating the long-term economic value of protection and conservation in comparison to competing resource uses.

Development and support for appropriate tourism activities is one of the ways that protected areas may be able to use to generate revenues and to demonstrate their wider economic contribution. The CBD Guidelines on Biodiversity and Tourism identify a number of potential benefits of tourism in protected areas that include:

- Revenue creation for the maintenance of natural resources of the area;
- Contributions to economic and social development, such as:
  - Funding the development of infrastructure and services;
  - Providing jobs;
  - Providing funds for development or maintenance of sustainable practices;
  - Providing alternative and supplementary ways for communities to receive revenue from biological diversity;
  - Generating incomes;
  - Education and empowerment;
  - An entry product that can have direct benefits for developing other related products at the site and regionally;
  - Tourist satisfaction and experience gained at tourist destination.

It may also be possible through tourism for protected areas to improve relationships with stakeholders, especially local communities, private sector and NGOs, and to gain enhanced recognition as part of the national heritage. The economic effects of tourism to protected areas can provide a rationale for continued investment in their protection, and stimulate general support for conservation. Furthermore, the 'user pays' approach offers a simple mechanism for raising funds through tourism, particularly with the present trend of growing interest amongst tourists in visits to protected areas.

However while tourism may seem an attractive option for some protected areas, a variety of factors can affect the development of tourism activities as a successful tool for revenue generation. Amongst these are the seasonality and volatility of tourism demand particularly in lesser-visited sites and in developing countries. Tourism levels can drop sharply where there are security concerns in any part of a country or region, even if a protected area is itself safe to visit. The sector is also highly competitive, and demand may drop as competing sites or destinations become available, or tourism fashions change. Thus the potential for tourism at a protected area is no guarantee of the level of tourism, or its stability, over time, as these are subject to many external influences beyond the control of any protected areas. Secondly, while the global scale of tourism is enormous, the range of travel and tourism possibilities from which tourists can select is also vast: sophisticated market-orientated product development will be necessary for a protected area to be successful in attracting a regular flow of tourists.

A third factor is whether the legal frameworks and administrative arrangements under which protected areas operate in a country, allow them to develop fund-raising activities to support their management and conservation programmes. The main organisational arrangements for protected area management and their implications for fund-raising through tourism.

The fourth factor relates to the ability of a protected area to manage tourism. While there are some excellent examples of protected areas that manage tourism well and raise considerable revenues from tourism, these are mostly located in developed countries and are easily accessible to large markets with affluent populations. For many protected areas, particularly in developing countries, their markets for tourism will be smaller and less secure, and the sites themselves underfunded. As a result, they may not have sufficient human or financial resources to cover the start-up or even running costs of tourism activities.

The fifth factor is the effect of tourism in a protected area on future tourism development that might affect that area. There are many examples around the world where small-scale tourism has stimulated further poorly controlled developments that result in both environmental damage and a decline in the value of an area for tourism. This can happen as much through uncontrolled developments by the informal sector often of bars, restaurants and cheap accommodation as through mass tourism developments. Therefore it is essential that a long-term approach is taken to tourism activities in protected areas, and consideration given to the ways in which developments can be controlled and limited to scales and types that are compatible with conservation; and to whether and how such controls and limits may be enforced effectively. This will generally require the existence and enforcement of effective planning laws, integration of tourism and conservation issues at protected areas with wider regional plans, and support from government authorities, local communities and the tourism sector.

Alongside this, there has been growing realisation that tourism even at quite low levels can have a number of adverse impacts on protected areas, and that these need to be managed. Adverse impacts include disturbance to wildlife, which can affect feeding and breeding success, damage to sites through trampling and erosion along footpaths and viewing sites, and impacts from solid and liquid wastes if these are not managed properly. Some species and habitats are more susceptible to these impacts than others and as a result tourism may alter the balance of biodiversity in protected areas. Management of these impacts requires staff to handle visitors and minimise visitor damage, investment in appropriate infrastructure such as erection of trails and interpretative signs, and provision of sanitary and waste management facilities, as well as planning for the management of tourism. Protected areas therefore require funds if they area to be able to manage tourism alongside their conservation goals, and it is reasonable to look for these funds from the tourism sector. It is also important to note that failure to manage and avoid or minimise adverse impacts from tourism also leads to costs to protected areas in the form of damage to the biodiversity that they have been established to protect.

Survival of Protected Areas

The survival of any protected area depends on the quality of its natural features, support from local communities and key stakeholders, political support through government policies and practices, and allocation of sufficient financial and human resources for effective conservation management, backed by clear plans and management systems for the site. These six survival essentials are interlinked, and for successful protection of sites, need to be mutually supportive: none are sufficient one their own to ensure the long-term survival of a protected area, and to some degree they are interlinked.

If there is good social and political support for a protected area, then it may be easier to obtain the financial and human resources required for its management. If a protected area can show not only that it protects high quality natural features, but that these features create economic value through tourism, as well as by providing environmental services, then political and social support for its survival is likely to be increased. It is important to recognise that the possibilities for protected areas to raise funds through tourism are just a part of one aspect of their survival. Even if these funds are not used for conservation purposes, but returned to central government, this can generate valuable political support for continued protection of the site. So too, if surrounding communities are benefiting from tourism generated around a protected area, then this can build local support for conservation, even if the income that the protected area receives from tourism is only small. In other words, tourism to protected areas generates a range of non-economic benefits that are of value to sites for their continued survival. These are difficult to quantify but may be of equal or even greater importance for a protected area than revenue generation, depending on its local relations and the way it is allowed to operate with the legal frameworks for protected area management that are set by the government.

Figure 1. Survival essentials for a protected area

It is therefore important for protected area managers to adopt broad approaches in site management that take into account the six survival essentials, and to strengthen capacity in each area. These are major constraints towards more effective approaches by protected areas to tourism and revenue raising consistent with conservation goals. The CBD's Programme of Work on Protected Areas also highlights the importance of multiple uses of protected areas, the need to improve site-based planning and management, and for active involvement of stakeholders in this. In order to prevent and mitigate negative impacts, the Programme calls for environmental impact assessments for plans and projects with potential to have effects on protected areas, and for timely information exchanges between all organisations and stakeholders that are involved. The six survival essentials for protected areas need to be kept in mind, which focuses on the possibilities for protected areas to raise funds from tourism.

Protected areas can be funded from a variety of sources, including government funding, multilateral and bilateral donor funding, donations from philanthropic foundations, corporations and individuals, as well as by raising revenues from people visiting or operating businesses associated with these sites. In addition there are sometimes opportunities to generate revenues through less traditional mechanisms, including through cause-related marketing· biodiversity prospecting, commercial and bilateral debt-for-nature swaps, trust funds and carbon offset projects.

Table 1. Potential sources of revenue in protected areas

- Government funding (mandatory or discretionary)
- Public investments
- Multilateral and bilateral donor funding,
- Donations from philanthropic foundations, corporations and individuals
- Revenue-raising methods:
  - Protected area entrance fees
  - Recreation service fees, special events and special services
  - Accommodation, transportation and guiding
  - Parking
  - Equipment rental
  - Food sales (restaurant and store)
  - Merchandise sales (equipment, clothing, souvenirs)
  - Licences, permits, and taxes
  - Licensing of intellectual property
  - Sale or rental of image rights (e.g. for taking photographs)
- Cross-product marketing
- Private sector initiatives
- Cause-related marketing
- Biodiversity prospecting
- Commercial and bilateral debt-for-nature swaps
- Trust funds
- Carbon offset projects

The potential contribution of tourism to the funding of protected areas needs to be set in the context of other funding sources available to protected areas. Each funding source brings with it a degree of risk, and as with any financial package it is important to balance the risks and opportunities presented by each one. The key is to build up a portfolio of revenue streams that are suited to the specifics of each particular protected area, and which together offer greater stability of funding flows than any one mechanism could provide on its own. The focus of any portfolio of revenue streams will inevitably depend to some degree on the features of different protected areas. For example, generation of revenues through fees, concessions and sales is most likely to be appropriate in protected areas where visitation levels are high.

Tourism Markets and Protected Areas

According to the World Travel and Tourism Council, the economic activity from travel and tourism will generate USD 5,4909 billion in 2004, supporting over 73 million jobs directly, and three times that amount indirectly. For the past decade, tourism has been growing at a global average of 4.5 percent, in real terms, each year, and this rate of growth is forecast to continue. Growth of tourism to some of the least developed countries is far higher than this average, and tourism is an increasingly important economic development tool for many developing countries. It requires lower capital expenditure for job creation than other industries, and generates employment particularly for women and young people, as well as providing opportunities for entrepreneurship and development of small firms. At the same time, however, tourism also has negative impacts. In particular, tourism tends to take place in the some of the world's most fragile of environments coastal zones and mountain regions and if not managed properly threatens the very resources that attract tourists in the first place, as well as causing damage to the ability of these regions to provide environmental services.

Forecasts for the expansion of international tourism show that some of the countries expecting to grow their travel and tourism demand the fastest in the world are also home to the key Global 200 Ecoregions that have been identified around the world by WWF. This suggests that there will be both a potential for protected areas to extract some revenues from the growth of tourism in these regions, and a likelihood that the growth of tourism will put increased pressure on some of the most important sites globally for conservation. It is important to recognise that for many of the countries where Global 200 ecoregions are located, tourism is regarded as a vital form of economic development. Most of these countries rely on tourism receipts to develop their economies while protecting their environmental resources and sustaining their social structures. Tourism comprises a variety of market segments. The main market segments that affect protected areas are mass tourism, adventure tourism and ecotourism/nature-based tourism.

Mass Tourism

Mass tourism is the largest component of the international tourism market. Mass tourists are looking for relaxing holidays often based around with 'sun, sea and sand' and entertainment. Most mass tourism holidays are sold to tourists as holiday packages that comprise flights, local transfers, accommodation and meals, and which are organised and marketed by tour operators. As part of their holidays, mass tourists may choose to go on one or more excursions to visit local attractions. These may be local towns, shopping areas, or theme parks, but also may include visits to famous sites, many of which may be in protected areas.

Mass tourists tend to be more likely to visit cultural sites rather than natural sites, but some excursions incorporate visits to natural sites to view wildlife, to enjoy the surroundings, or to participate in hiking, boat trips, snorkelling or scuba diving. Within the mass tourism segment, there is a growing demand for excursions, and consequently where protected areas located near to mass tourism resorts, there may be opportunities to work with the tour operators to develop excursions that visit linked to protected areas.

Adventure Tourism

Adventure tourism is a growing segment of the speciality tourism market, and involves strenuous outdoor activities, and in some cases potentially dangerous activities such as mountain biking, trekking, white-water rafting, or paragliding. This can involve travel in remote areas, some of which may be protected. The prime motivation for travel is adventure rather than to enjoy nature.

Many outdoor activities engaged in by adventure travellers are potentially damaging to site conservation. For example, mountain biking can cause serious damage to paths and trails in fragile habitats, as well as general disturbance to wildlife. It is therefore important for protected areas to manage any adventure tourism activities to keep them away from the most sensitive parts of sites, and to limit the amount and timing of adventure activities; and to raise awareness amongst adventure travellers of ways to minimise their impacts and of the main conservation activities on the site.

Ecotourism/Nature-based Tourism

The primary objective of ecotourism/nature-based tourism is to visit and see attractive natural environments and their wildlife. Examples are birdwatching, whale watching, game viewing, scuba diving, botanical tours and nature photography. According to World Tourism Organisation (WTO) surveys in key tourism generating markets, ecotourism enthusiasts are mostly people from relatively high social brackets, high levels of education, over 35 and women slightly outnumber men.

Ecotourism and nature-based tourism in general have been tipped as the key segments of the tourism sector that will generate and spread benefits into conservation. However, in order to gain any benefits from this type of tourism, it is necessary for policy makers, conservation managers and protected area administrators to understand these tourism markets and how to use these to attract tourists.

Managing Tourism in Protected Areas

While tourism can be a source of benefits for protected areas, in many cases, protected areas may not have the resources or access to investment that is needed to turn these potential benefits into a reality, and may not be sufficiently equipped to control and manage tourism so that it remains in balance with conservation goals. Additionally, any adverse impacts that are caused as a result of tourism are a cost to protected areas. The first priority for protected areas is therefore to find ways of working with the tourism sector to reduce the impacts of tourism and costs to a site of managing tourism, before exploring the potential of using tourism to raise revenues that can contribute to protected area management.

To help protected areas and national authorities address such concerns, the Convention on Biological Diversity has adopted a set of Guidelines on Biodiversity and Tourism· These are designed to provide a framework for the management of tourism within protected areas, consistent with the conservation and sustainable use of biodiversity. The CBD Guidelines on Biodiversity and Tourism address market issues and tourism trends at all levels, as well as key factors for management of tourism in protected areas, such as establishing limits of acceptable change, zoning and control of tourism, impact management measures, and promotion of responsible behaviours by tourists visiting protected areas.

The Guidelines are applicable to all forms of tourism and tourism activities, and stress the importance of effective planning controls on tourism development, and of public education, capacity building and awareness raising concerning tourism and biodiversity. They also stress the importance of collaboration between protected areas, the tourism sector, indigenous and local communities, national and local government authorities, and NGOs, in order to plan and manage tourism. In particular, the World Commission on Protected Areas notes that : "The operation of a protected area tourism industry requires the cooperation of both the public and private sector. Neither can do the job alone. Each is fundamentally dependent on the other." Protected areas provide opportunities for tourism, while the commercial tourism sector provides the opportunities and services through accommodation, catering and transport, as well as marketing for tourists to visit protected areas. By combining the entrepreneurial skills and links to tourism markets that tourism businesses have, with the conservation skills of protected area managers, it is possible to provide a better experience for tourists, and to gain a better contribution from tourism for protected area conservation.

Successful tourism in protected areas requires the ability to develop and market tourism products based on protected areas, and the ability to maintain the quality of these areas for the future. The tourism potential of any protected area depends on a wide variety of factors, that include location, accessibility, market demand, proximity to popular tourism destinations, publicity, presence of local tourism businesses and infrastructure. The ability of a protected area to manage tourism depends on the implementation of clear management strategies, the scale of demand for visits to the site, the staff and resources available for management of tourism, and the legal and political environment covering wildlife protection in the countries in which they are located. These factors vary considerably amongst protected areas, and it is therefore important for tourism and conservation strategies to be developed for each protected areas, so that they can be tailored to their specific circumstances.

Simple Economic Model of Tourism and Protected Areas

Figure 2 shows a simplified model of the monetary flows associated with tourism and protected areas. Protected areas are able to receive funds from three main sources: through government funding, by charging entrance fees to visitors, and by charging businesses that operate within the boundaries of protected areas. At the same time, funds flow out of protected areas in several ways, including via payment of staff costs, procurement of the goods and services that protected areas require in order to operate, and as a result of damage that may be caused to a protected area.

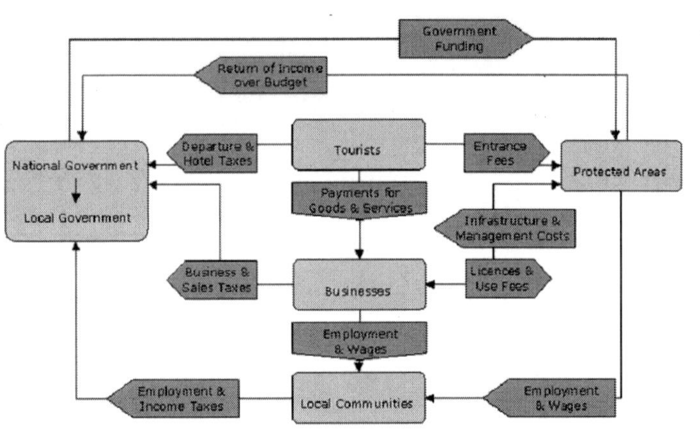

Figure 2. An economic model of tourism in protected areas

This model is subject to several considerations. First of all, where protected areas are run as a part of government public services, they may be subject to strict budgetary limitations and rules. These often include limits on the amount of funds that protected areas are allowed to raise by charging for access and services, and may also include requirements that any additional funds raised by protected areas be returned to central government. Such limits and requirements may be set out in legislation or government regulations, and if so, are likely to be difficult to change, particularly if changes have to be debated and approved by state or regional legislatures.

Secondly, where protected areas are able to charge tourists and businesses for access to and use of these sites, there are costs involved in collecting the charges and in providing the necessary infrastructure even if it is only basic and in managing tourism in protected areas. It is important to remember that for tourism to provide net financial benefit for the management of a protected area, the inflows of funds from tourism will need to exceed the costs that are incurred by protected areas in providing necessary infrastructure and in managing tourism. If the tourism-associated costs incurred by a protected area are greater than the inflow of funds it receives from tourism, then that protected area is essentially subsidising tourism.

This basic economic picture holds true in all circumstances, even though protected areas may vary considerably in their ability to control tourism in and around their sites. For some protected areas, their geographical remoteness may keep visitor numbers low, and well below levels that would start to cause problems. But where protected areas are located close to centres of population and tourism resorts, visitor numbers are likely to be high and difficult to control. While the basic economic picture is the same in each case, these protected areas differing only in location and accessibility to tourists would need to use very different strategies for management of tourism. In the case of remote protected areas, there may be a desire to increase the number of tourists within defined limits, in order to raise additional funds.

In the case of heavily visited protected areas where tourism is difficult to control, a main focus may be to find ways of reducing adverse impacts from tourism so as to reduce the drain on funds, rather than to raise additional funds. In a few cases, groups of protected areas with varying levels of visitation have been formed in an effort to reduce visitation at the most popular sites, by encouraging tourists to visit other sites in the group: a proportion of tourism revenues may also be pooled for use by less visited sites.

The economics of protected areas are also closely linked to political factors. At a local level, communities in and around protected areas, particularly where populations are growing, may seek access to these areas to support their livelihoods through agriculture and resource extraction. To counter these pressures, it is important for protected areas to demonstrate an economic value to local communities, for example, by employing community members to assist with conservation and management, and by supporting development of alternative livelihood options, of which tourism may be one, which are less damaging and have potential to improve local incomes. In the absence of employment and alternative livelihood options, local communities are more likely to use protected areas illegally and in damaging ways, including small scale logging and resource extraction, land clearance and settlement.

At national level, governments regularly come under pressure from developers and major investors to open up protected areas for a variety of purposes, including timber, oil and mineral extraction, and development of tourism resorts. Unless governments can clearly see and understand the long-term benefits that protected areas provide, it is likely that they will support schemes involving inappropriate activities and development in protected areas. By showing that they have an economic as well as symbolic value when they are managed sustainably, and by building local level support, protected areas will be in a stronger position to counter pressures for major developments and resource extraction. This can include showing that the viability of national tourism sectors, and the revenue that these can provide for national governments, is linked in part to the existence of well-managed protected areas which are valued by international tourists as well as by local communities.

These political factors mean that it may be appropriate for protected areas to encourage and support suitable and controlled tourism activities for the political benefits they can generate, even if they do not provide net financial benefits.

In reality, few of these potential sources are actually exploited within the protected area system of any one country, highlighting the potential for protected areas managers to increase their income by utilising a much broader range of revenue management options. To take advantage of these opportunities, protected area systems need to build capacity in a variety of ways. Factors crucial to building a financially sustainable system include skilled personnel who can analyse financial needs and opportunities, and select approaches appropriate to each area; infrastructure for the management and visitor services needed, including accommodation, communications, and transportation; an enabling policy environment in which necessary actions (such as dedicating revenues to the system) can be accomplished; clarity over where and how fees are used; and developing systems for community participation.

Structures for Protected Area Management

Directly or indirectly, government departments have responsibilities for the administration of the majority of protected areas. However, factors such as shortages of government funds, lack of sufficient trained staff, and demands within governments and by donors for more efficient and 'businesslike' management of protected areas, for their greater accountability, and for greater participation of key stakeholders including local communities have led to the emergence of a variety of other approaches to the management of protected areas.These include management by:

— government departments or agencies

— parastatal bodies

— NGOs

— community organisations

— the private sector

— a combination of these.

Government Departments

The most common form of protected area management is management by government departments. Protected areas are generally administered through Ministries of Environment, or Natural Resources, and frequently through specific sub-departments of these ministries. In systems of federal government, responsibilities for protected areas may be allocated to either state or federal governments, or split between the two levels of government. The local government level may sometimes also have some limited responsibilities for protected area management.

Funds for protected areas managed by government departments depend almost entirely on allocations made by government from the national budget. In most developing countries where there are severe budgetary constraints, the amount of government funding available for protected area management is likely to be very low. Furthermore, some countries have legal restrictions that prevent government departments from generating their own incomes for example, by selling concessions to businesses and / or which require any revenues raised by government departments to be returned to the national exchequer. Restrictions of this type may prevent protected areas from supplementing their government funding with other sources of income.

In consequence, protected areas that are managed through government departments are often underfunded. In addition, protected areas managed in this way may be subject to changes and uncertainty associated with political matters at national level for example, where protected area managers and policies are changed following changes of government and as a result may not be responsive to local issues and the concerns of communities located in or around protected areas. Local communities are often concerned that they see little if any of the benefits generated by protected areas managed by government departments, while experiencing disadvantages from the presence of protected areas locally.

Parastatals

A parastatal is essentially a public sector organisation with some elements commonly found in private sector organisations. The main advantages of parastatals include their flexibility to set fees and charges, establish funding mechanisms such as concessions, and respond to customer needs; their ability to retain the money they earn which gives a resulting incentive to generate funds through greater entrepreneurship; and their freedom to implement staffing policies based on efficiency and market salaries.

Parastatals are also less tied to political changes than government departments, and are able to operate more independently as a result. This is particularly important to their success, as they are less affected by changes of government policy and priorities

Parastatal agencies have been established in countries where institutional arrangements inhibit protected areas from themselves providing the diversity and quality of tourism opportunities from which revenue flows back into the protected area. In various countries, including Kenya, Tanzania, South Africa, Indonesia and some Caribbean countries, the response has been to establish parastatal agencies, which function like companies within the public sector, and which operate at 'arms length' from governments. Some African parastatal agencies were found to have some 15 times as much funding as did government run protected area authorities· In Indonesia, the emergence of parastatals is in keeping with moves towards decentralisation that has been taking place since the end of a heavily centralised political regime in 1998.

Ideally, parastatal agencies should be managed by a Board of Directors who are experts in various aspects of tourism, business and protected area management. These members create vital links to the commercial world, thereby enhancing the potential financial gain by the protected areas themselves. However, parastatals do not have as much autonomy and flexibility as true private sector organisations because they are responsible for public resources and therefore remain accountable to the public. They are unable to make decisions strictly on the basis of business criteria because they need to maintain non-profitable protected areas as well as the profitable ones, and they should provide benefits to local communities. Their Boards are usually appointed by the government and may be based on political or other vested interests rather than ability, commitment, etc. The public will often continue to regard them as public sector organisations and may object to their paying higher salaries, entering into business contracts, deviating from government procurement processes, etc.

Private Sector

In a small number of cases, governments have transferred management of specific protected areas to companies in the private sector. Transfer agreements between the governments and private companies concerned, specify the obligations of private company managers to ensure protection of the site, while providing flexibility to allow companies to develop revenue streams that can be used to fund protected area management and to generate profits generally, through wildlife-related tourism, and establishment of tourist lodges.

In addition, a number of protected areas have been established by privately-owned companies on privately-owned land, particularly in Southern Africa. These protected areas operate on a commercial basis, and generate funds through tourism. Several have been established on land of previously limited wildlife interest for example, on farmlands and have built up good wildlife populations through reintroductions, breeding programmes, and effective conservation management.

Community Organisations

In community-based management of protected areas, management is transferred to one or more communities with a recognised association with the protected area. The communities to which management is transferred are identified as having either current or previous ownership or settlement of the protected areas concerned, or usage rights over those areas.

Community-based management schemes have advantages in that the benefits resulting from sustainable management of protected areas can be shared by the communities, and these communities are able to manage their use of the land, its wildlife and natural resources in an integrated way. This therefore avoids conflicts that can often arise between communities and protected area managers under other forms of protected area management.

Against this, communities will generally lack the skills and capacities that are required for successful conservation and management of protected areas, and depending on the specific ways in which individual community-based management schemes are set up, may be free to undertake a wide range of potential activities on the land concerned, of which conservation and protection is just one option. They are also unlikely to have the skills to develop and market more than basic types of tourism.

These aspects can generally be addressed by providing technical and capacity building support to communities involved in community-based management of protected areas, by setting legal limits on what uses and types of management are allowed in the areas concerned when community-based management schemes are set up, and by negotiating management agreements with communities under which management of some areas is leased to other organisations which pay fees for the right to use specific areas under conditions that are set out in a contractual agreements with the communities concerned.

For community-based management of protected areas to succeed, the following elements are necessary:

— External support, capacity building and awareness raising provided by protected area administrations and managers
— Demonstration that the economic benefits that communities can derive from conservation are at least as good as those obtainable from other uses, both for the community as a whole.
— Integration of community requirements into the protected area management approach.
— Realistic development and pricing of revenue raising activities.
— Effective decisionmaking and management structures within communities, with decisions being taken democratically with the participation of all community stakeholders, and transparency and accountability in the management of revenues derived from protected areas

NGOs

In some cases governments have transferred responsibilities for protected area management to suitably qualified and experienced NGOs. As with transfers of protected area management to the private sector, such transfer agreements between governments and NGOs specify the obligations of the NGO managers to ensure protection of the site, while providing flexibility to allow NGOs to develop revenue streams that can be used to fund protected area generally, through wildlife-related tourism, or sale of concessions to private businesses.

If management is transferred to a large international NGO, then that NGO may be able to draw on fees and donations from its members to help support protected area management. NGOs may also be able to attract donations from visitors to the sites that they manage. However, the financial security of arrangements for management of protected areas by NGOs depends on the size and management of the NGOs involved. NGO management of protected areas can potentially operate in a wide range of settings, including in areas where there are limited possibilities for development of revenues streams from tourism or other commercial activities, and where management by private companies would therefore not be possible. However, delegation of management of protected areas to NGOs may be a sensitive issue both for governments particularly where the NGO concerned is foreign and for local communities who may regard management by NGOs as more distant and less desirable than management by government agencies.

Hybrid Arrangements & Partnerships

In many cases, management of protected areas involves balancing of a wide range of interests, including the legal responsibilities of government departments and agencies for planning and protection of sites, the interests of local communities and the private sector, and of NGOs concerned with conservation. None of the structures outlined above incorporates all these interests, or is able on its own to encompass the full range of expertise that is available across these different types of organisations. To address this, some protected areas have developed structures such as advisory boards and liaison committees that are designed to bring together these different interest groups, and to provide a channel for them to input to plans and decisions on protected area management. These boards or committees may be established on an informal basis, or through more formally constituted arrangements, some of which may involve government approval or legislative changes: they provide complementary structures to assist site management. Their benefits include bringing additional expertise and capacity into protected area management, establishing channels to raise funds that are returned for use in conservation, and for building better links with local communities and businesses.

Hybrid arrangements and partnerships also provide important routes for raising funds for protected areas, particularly where protected areas are prevent from raising significant revenues themselves for example, because of restrictions on public sector bodies, and/or because of limited revenue raising opportunities. Fund-raising partnerships are generally based around non-profit Foundations or NGOs, which are able to raise funds by various means, including through donations and, in the case of NGOs, some business activities such as sale of products and services. Fund-raising activities are generally explicitly linked to the protected areas where funds will be used to support conservation activities, and the Foundations or NGOs involved will generally have some form of agreement with the protected areas concerned, over both use of funds raised, and use of the names and logos of protected areas for fund raising purposes.

Some protected areas have themselves joined together to pool their resources to market the tourism they offer, and through this to try to direct tourism to lesser visited sites and away from areas that are already saturated by visitors. Some heavily visited sites may also allocate a portion of the revenues generated by tourism to assist conservation and tourism projects at less visited sites.

Funding Methods for Protected Areas

Protected areas that plan to raise funds from tourism will need to select and implement suitable funding raising mechanisms for this. The main mechanisms that protected areas use to raise funds from tourism are covered:

- Entrance fees
- User fees
- Concessions and leases
- Direct operation of commercial activities
- Taxes
- Volunteers and donations

The suitability of these mechanisms for any particular site will depend on a variety of factors, including the scale of tourism at the site, where tourism takes place within the site, access points, and the way in which the commercial tourism sector interacts with the site.

Entrance Fees

Setting an appropriate protected area entrance fee one that covers the protected area's capital costs and operating costs, and ideally even the indirect costs of ecological damage is one of the best and most used ways for management agencies to capture a larger share of the economic value of tourism in protected areas and protected areas. One in two protected areas use some method of fee system, but few currently generate substantial revenues from this, and generally the charges do not reflect the service and product offered by the protected area. There has been reluctance to charge the full cost of running the protected area for various reasons: traditionally, protected areas have been funded from the national treasury as a public good; a collection and accounting mechanism may be difficult and expensive to set up; and there may be a fear of resistance from users.

However, studies have shown that visitors are willing to pay more if they know their money will be used to enhance their experience or conserve the special area they are visiting, and that visitors are often willing to pay higher entrance fees than those currently charged, particularly for protected areas with a high level of demand.

Entry fees are sometimes literally collected at the point of entry to the protected area, with a set amount per car or per individual. In some cases 'point of entry' collection may be difficult because of the size of the protected area or if there are multiple entry points. In such cases, it will be easier and more effective to levy user-fees at the point of activity, for example with diving or car-parking. The entrance fee may include an immediate tangible 'good' in the form of a map and/or information on the protected area.

Establishing a policy on entry fees begins by defining the purposes of the fee program. Pricing policies are generally set at national level. In some countries, it may be decided that each protected area should then use these policies to set their own prices, while in other countries centralised governmental control may be necessary to provide an overview of relative pricing between protected areas at national level or so that, for example, it would be possible to implement a visitors' pass system for access to several protected areas.

Many protected areas have begun with a single level of admission fee and gradually developed a differential or tiered pricing system. This more sophisticated pricing approach allows protected areas to charge visitors according to a number of factors including the time of year, the level of service provided and the income or place of origin of the visitor. For example, a lower price may be charged to domestic visitors to encourage their interest in conservation, and to reflect the ability of domestic visitors to pay, while enabling higher fees to be charged to international tourists from high income countries. Many protected areas now adopt such a system for instance, foreign visitors to Ras Mohammed, in Egypt, pay $5 while Egyptians pay $1.20. Other tiers are possible, for instance allowing lower fees for students, senior citizens or groups. Differential pricing however is anything but straightforward.

In theory, admission fees can be used as a tool to keep the number of visitors within an ecosystem's carrying capacity or to limit growth rates, so that planning, management and control are not outpaced by development. In practice, developing country protected area authorities in particular may lack the skills to assess and manage tourism flows in this manner, particularly if there is pressure from the private sector to allow increased visitation.

Challenges to successful implementation of entrance fees include inefficient fee collection, which results in losses of entrance fee revenue from protected areas. There is also a risk that scarce personnel resources will be redeployed towards collection of fees rather than protection of resources. In addition, corruption and bribery at entrance gates are common features. This can be avoided by a good and transparent system of accounting for revenues and expenditure, or by designing a cash-free system of entry at gates. For example, in an attempt to overcome corruption, the Kenya Wildlife Service (KWS) has introduced an electronic debit card system for paying for entrance to a protected area. Visitors charge their card by paying cash at KWS headquarters in Nairobi where it is efficiently collected and accounted for. Then, on visiting the protected areas the cards are 'swiped' in machines at the gate that reduce the credit on the card by the amount of the entrance fee. This system seems to be working well to increase overall takings.

User Fees

Iin some cases it will be difficult to collect a literal 'entry fee' due to geographical factors or because it is more appropriate to levy management and conservation contributions from users of facilities within a protected area. User fees may be charged for use of facilities such as car-parks, campsites, visitor centres, mountain huts, or canopy walkways, or for carrying out activities in the protected area such as fishing, hiking or mountain climbing, sailing, or hunting. Two of the most lucrative forms of user fees are for scuba diving and trophy hunting. It is important to encourage tourists to be willing to ay user fees, particularly by showing how the fees are used to support conservation and management in protected areas.

In many countries, such as Indonesia, dive fees have generally been developed on a case by case basis, rather than through central planning at national level. Some parks in Bahamas, Netherlands Antilles, Mexico, Venezuela and Belize charge divers while other national parks in same countries do not.

Reluctance to implement fees may be due to the cost of introducing and maintaining a fee collection system and, in some areas, political or socioeconomic factors. However, dive tourism is growing and in many of the more popular places the theoretical recreational carrying capacity of MPAs is being reached, while many suffer from a shortfall in financial resources. There is therefore a strong case for converting diver willingness to pay into higher fees that could make a significant contribution to conservation of marine protected areas.

Concessions & Leases

Traditionally, protected area attitudes to the private sector encompassed a range of approaches including trying to keep them out or ignoring them, marketing to them, licensing or restricting them, competing with them, and forming partnerships with them. However, in the current climate of shrinking funds for management of protected areas, coupled with increased awareness of the value of managed market forces, there is now a trend towards greater cooperation with the private sector. Governments increasingly recognise the value of providing an enabling environment for the private sector to operate within protected areas.

Around one in five protected areas use some form of concession or licensing system, and the increasing privatisation of service delivery in protected areas will increase their popularity. Depending on the legal framework of the country, any function or privilege of the state from operation of individual facilities to management of the entire protected area can be contracted to a concessionaire. Concessions and leases cover a range of permits, leases and licences. Common features are that they allow private companies or individuals to run commercial operations within a protected area while generating financial benefits for the protected area. Activities may include tour guiding, trekking or diving operations, accommodation provision, restaurants, souvenir shops, sport fishing or hunting trips, horse-trekking, hire of kayaks or mountain bikes and the hire or sale of other sports and recreational equipment.

Concessions or licenses can be granted to private companies, community groups, or NGOs or other not-for-profit enterprises, and can offer a way of ensuring that communities living in and around the protected area can benefit from it economically.

Concessions management necessitates a balance between resource protection and visitor needs. Government policy and regulations may require that management of visitor services provided by concessionaries and leaseholders take into account public opinion, provide employment for local people and other elements of benefit to local communities, and include regular monitoring of environmental impacts.

Concessions can make a considerable contribution to management of the protected area. For instance, seven concessions awarded to private consortia in South Africa in 2000 guaranteed SAN Parks at least Rs 202 million (over USD 30m at 2004 rates) over a twenty-year period. An important element of concessions is that, where local communities are involved, the link between tourism earnings and conservation of the protected area must be made very strongly. Where this is not done, local people will be much less inclined to protect the protected area. For instance, a study in Komodo National Park, in Indonesia, found that although residents had positive attitudes towards tourism and that there was support for conservation, there was no positive connection between the two. Tour operators have a very strong role to play in communicating conservation messages.

Direct Operation of Commercial Activities

An alternative to the system of private concessions, is for protected areas management bodies to generate income through operating services themselves. These may include any of the services traditionally handed over to private concessionaries because of the lack of skills within protected areas services such as accommodation, guiding, equipment hire, or sales of merchandise. With the increasing sophistication of the tourism industry and, in many countries, increased ease of finding skilled personnel, along with an awareness of the contribution to protected area revenues which tourism can make, protected areas administrations may more easily be able to become involved in such commercial activities in their own right in order to maximise protected area revenue. This can be done either directly, if national legislation systems allow, or indirectly through a state-owned company. A suitable format might also be some kind of public-private sector partnership such as a joint venture with a private company or with local communities. This kind of arrangement would help to stem the siphoning-off of funds through over-allocation of private concessions, and in the case of local communities could help them with marketing and capacity-building.

Strong management is needed to prevent protected areas personnel simply setting up private enterprises on their own behalf, such as at Bromo Tengger Semeru National Park, in Indonesia, where rangers are involved in a range of lucrative activities from hotels to internet cafes, with no financial benefit to the protected area. Another possibility for generating income arises when protected areas are used as sets for filming, subject to payment of appropriate fees. Books written by protected areas personnel can provide another source of income in the royalties generated, although this is never likely to be large.

Taxes

A further method of financing protected areas is through taxes. These may take the form of national taxes levied on all visitors to the country or on users of particular tourism services or products, local taxes levied on users of the protected area or on the use of equipment (it will be seen below in the examples how these differ from user fees). The revenues raised are then used for conservation. The advantages of using the tax system include the ability to generate funds nationally (or regionally) and on a long-term basis; the freedom to use funds to suit a variety of needs, as accountability is to the public at large and not to a specific donor; the possibility of using such funds as a "matched" component of funding from international donors, who are increasingly requiring evidence of national commitment as a prerequisite for support; and ease of collection, since there is usually no need to set up a new collection system.

There are also disadvantages of the system. One is that it can be seen as less fair than collecting fees directly from protected areas users, as all visitors to the country/region are taxed for services/resources they might not use. Other difficulties may arise in winning political support for new taxes and setting them aside for conservation, particularly in countries where conservation is a low priority. In Belize, years of negotiations were required before a conservation tax was established, and it was set at a much lower level than originally anticipated. Here, as in many countries, special legislation was required for the tax to be earmarked for a special fund rather than paid into the general treasury. It may be necessary, for political reasons, to allocate revenue raised to different departments or activities. Furthermore, the upfront costs of lobbying for and building such systems also need to be weighed against their potential benefits, although the process of building a constituency to support protected areas through tax legislation goes hand in hand with constituency building for many other purposes—a necessary investment whether or not it leads to specific taxes linked to conservation.

A further problem with taxes may arise when there are too many small taxes, as these can be inefficient to collect and administer, easy to avoid, and may be an irritant to tourists if they have to pay each individually, or are inconvenienced by the payment process and associated bureaucracy. It is therefore advisable to simplify tourism taxes, and where possible, to provide mechanisms for tour operators to pay taxes directly on behalf of their clients, and to incorporate them into the overall prices of the tours that they offer. Bed levies are common around the world, and this is most effective when the area is within one municipality or protected area. For example, in the USA, the state of Delaware imposes an 8% charge on room prices of which 10% goes to finance beach conservation. In the Turks and Caicos Islands 1% of the room tax goes to a protected area conservation trust fund (the total tax is 9%).

It can be argued that because tourism is a major source of environmental impact, including climate change, the cost of reducing or rectifying environmental impacts should be included in the cost of travel by air, for instance by adding a tax to the cost of a holiday or by taxing airlines on fuel consumption. Ring-fencing some of these funds for conservation would create a major source of income. However, strong economic and political lobbies oppose this, and air travel is beyond the scope of the Kyoto agreement on climate change.

Volunteers and Donations

Some protected areas have a policy of involving volunteers in their work, either through providing guiding and interpretation services, fund-raising or through staffing key services such as entry booths. This is likely to work best in industrialised countries where there are pools of relatively wealthy people with considerable disposable time. Examples of volunteer systems are those used in Canada· and in the Brisbane Forest Park in Australia. A further way of generating funds for conservation areas through tourism is via donations by tourists who have been to the area or have some interest in it and by private companies keen to demonstrate their CSR (Corporate Social Responsibility) credentials. This is a useful way of transferring funds from richer Northern communities concerned about conservation of the world's natural resources to poorer Southern communities who lack the finances to conserve these resources themselves. There are various ways of operating these schemes, including through trust funds and 'Friends of...' organisations. An additional conservation benefit of such groups is that subscribers can be encouraged to lobby against specific threats to the conservation of the area they are interested in.

Many tourism companies set aside a small proportion of their profits (generally less than one percent), or make a donation per booking, to support a variety of philanthropic activities, although these mostly focus on humanitarian rather than environmental organisations. However, tourism companies that specialise in nature-based tours, and some mass tourism companies, support environmental projects and organisations in the destinations visited, particularly if these improve opportunities for tourism, for example, by developing trails or interpretation and guiding, and appeal to their clients. Some companies also encourage their clients to make donations to the projects that they visit, generally by pointing out the work and funding needs of projects during the course of guided tours, and leaving it up to clients to donate if they wish. Sites can also help to encourage donations by setting up donation boxes in suitable locations, such as at visitor centres or in onsite accommodation or food outlets.

Trends in Park Tourism

Globally, the area of land covered by the world's parks and protected areas increased considerably from 1900 to 1996. By 1996 the world's network of 30,361 parks covered an area of 13,245,527 sq km, representing 8.84% of the total land area of the planet. Protected areas have been created in 225 countries and dependent territories. Figure 1 shows the growth of this network over a 100-year period. The impressive growth of the world's park network is the result of the widespread acceptance of the ecological ethic and aggressive political action. It appears that the tourism activity occurring at these sites created a self-perpetuating phenomenon of visitation, education, and desire for more parks, visitation, and education.

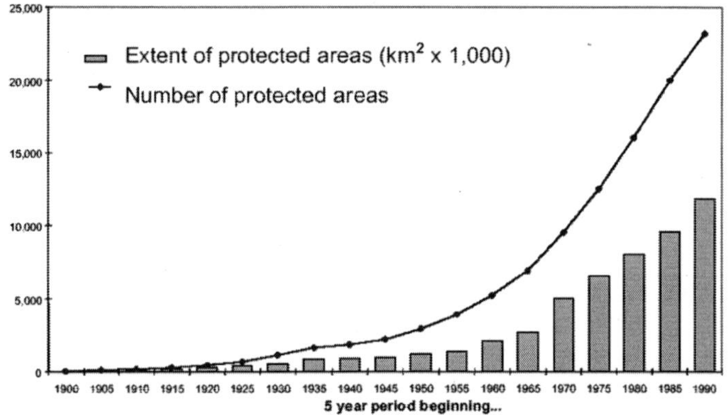

Figure 1. Cumulative growth of protected areas, 1900–1995.

The global network includes a wide variety of types of protected areas, ranging from nature reserves through to protected landscapes and managed resource protection areas, within IUCN's six-category system (Table 1). This facilitates accounting and monitoring at national, regional, and international levels. All six categories are well represented in the network, but with national parks and resource management areas being the two categories with highest representation.

The name "national park" is closely associated with nature-based tourism, being a symbol of a highquality natural environment with a well-designed tourist infrastructure. Eagles and Wind found that Canadian ecotourism companies frequently used the name "national park" as a brand name to attract potential ecotourists to their offerings. With 30,361 parks in the world, and with 3,386 having the well-known title of a national park, it is clear that any particular political unit, such as one country or one province within a country, has a major task to get its sites recognised globally. There are many sites available for tourists. Some countries, such as Canada, have the disadvantage of having many of their sites known as provincial parks, a name unknown outside Canada, and one which is suggestive of a lower level of importance.

Table 1. IUCN categories of protected areas

Category I	Strict Nature Reserve/Wilderness Area: protected area managed mainly for science or wilderness protection
Category Ia	Strict Nature Reserve: protected area managed mainly for science *Definition:* Area of land and/or sea possessing some outstanding or representative ecosystems, geological or physiological features and/or species, available primarily for scientific research and/or environmental monitoring.
Category II	National Park: protected area managed mainly for ecosystem protection and recreation *Definition:* Natural area of land and/or sea, designated to a) protect the ecological integrity of one or more ecosystems for present and future generations, b) exclude exploitation or occupation inimical to the purposes of designation of the area, and c) provide a foundation for spiritual, scientific, educational, recreational and visitor opportunities, all of which must be environmentally and culturally compatible.
Category III	Natural Monument: protected area managed mainly for conservation of specific natural features *Definition:* Area containing one, or more, specific natural or natural/ cultural features which is of outstanding or unique value because of its inherent rarity, representative or aesthetic qualities or cultural significance.
Category IV	Habitat/Species Management Area: protected area managed mainly for conservation through management intervention *Definition:* Area of land and/or sea subject to active intervention for management purposes so as to ensure the maintenance of habitats and/ or to meet the requirements of specific species.
Category V	Protected Landscape/Seascape: protected area managed mainly for landscape/seascape conservation and recreation *Definition:* Area of land, with coast and sea as appropriate, where the interaction of people and nature over time has produced an area of distinct character with significant aesthetic, ecological and/or cultural value, and often with high biological diversity. Safeguarding the integrity of this traditional interaction is vital to the protection, maintenance and evolution of such an area.
Category VI	Managed Resource Protected Area: protected area managed mainly for the sustainable use of natural ecosystems *Definition:* Area containing predominantly unmodified natural systems, managed to ensure long- term protection and maintenance of biological diversity, while providing at the same time a sustainable flow of natural products and services to meet community needs.

Unfortunately, there is no global tabulation of park usage, as there is for park area. However, individual country reports and personal communication with many scholars and park managers suggest that park tourism volume has increased considerably over the last 20 years. Figure 2 shows the recent trend from Costa Rica's national parks: a curve showing increases over time. The one period of decline was due to a weak economy in the USA, causing lowered travel to Costa Rica, combined with an 800% increase in park entrance fees for foreigners. The visitation growth resumed as the economy improved, a more suitable pricing policy developed, and the market accepted the increased fees. The growth over 20 years shown in Costa Rica is representative of parkuse growth in many countries. Differences in various countries would largely reflect with the speed of the visitor- use growth, not in the overall trend of increases over time.

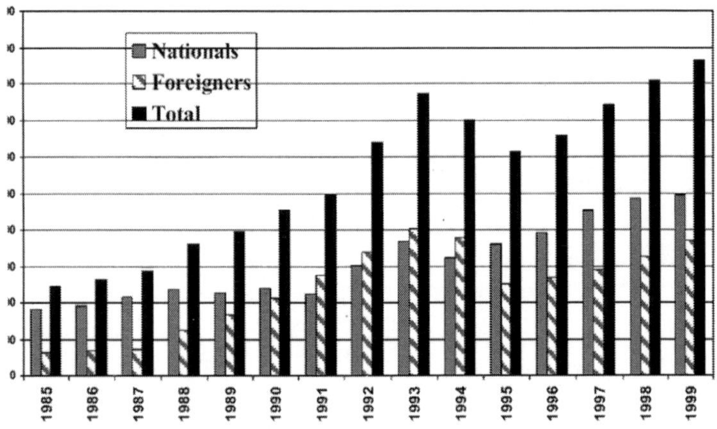

Figure 2. Visitation to national parks in Costa Rica, 1985–1999.

Eagles, McLean, and Stabler calculated the total national and provincial/ state park usage in the USA and Canada. In 1996 there was an estimated 2,621,777,237 visitor days of recreation activity in the parks and protected areas in these two countries. Clearly, the outdoor recreation occurring in the parks and protected areas in Canada and the United States is a very large and impressive activity. The 2.6 billion visitor-days of use per year represent major economic, social, and environmental impacts on society. One of the problems limiting international comparisons is the lack of accepted standards for park tourism statistics. International standards for park tourism data collection and management and global tabulation of these data are essential for comparisons to be made. In work done for IUCN's World Commission on Protected Areas, Hornback and Eagles outlined a structure and methodology for park visitation-use measurement and reporting. This approach has now been well accepted internationally, with many countries adopting the recommendations within their programs for measuring park tourism statistics. It is to be hoped that when the next version of the *United Nations List of National Parks and Protected Areas* is compiled, park visitor-use statistics, using the new standards, will be included in the global data collection effort.

Economics is an important component of societal decisionmaking, but it is often given low priority in the parks world. In parks the strong emphasis given to ecology is seen by many proponents as sufficient justification for public policy action. However, nature tourism is increasingly becoming important within sustainable development because of the potential of contributing to local and national economic development while also providing incentive for the conservation of biodiversity and nature in general.

Most of the world's protected areas charge low entry and use fees. These fees typically cover only a portion of the cost of protecting the resource and providing the features on which the park visitation depends. This pricing policy developed during a period where resource protection, a public objective that benefits all of society, was seen as the overwhelmingly important objective. If a public good benefits all, it is argued, it should be paid for by taxes on society. However, this logic falters when applied to outdoor recreation in parks, as only those who participate in outdoor recreation are beneficiaries. In a time of increasing parkland creation and widespread government financial retrenchment, it is increasingly difficult to justify continued increases in public expenditures to manage the parkland and subsidise the recreation of one segment of the population. Governments around the world are using this logic, in part, for the limiting the grants for park management. A reduction of budgets in the 1990s was documented for Canada and the USA, as was the development of new forms of park administration and new pricing policies. To address this issue, Parks Canada was reorganised in the mid- 1990s into an agency with a much stronger tourism focus and new business policy and focus. The business plan summarises the financial concept underlying the new agency with the statement that "subsidies will be phased out on services of benefit to individuals, by transferring the operation to the non-profit voluntary or private sectors, or these services will be stabilised on a full cost recovery basis"

There are dramatic differences among the world's parks in terms of pricing policy, tourism income, and financial management. A global study of biosphere reserves found that only 32 of 78 responding sites charged admission fees to visitors. The fees ranged from less than US$5 to $110 per person per day, with the vast majority at the lower range. There was a statistically significant relationship between total direct income and the numbers of visitors for all biosphere reserves. Higher visitor numbers corresponded to higher budgets. The authors concluded that "better financed biosphere reserves are likely to be better managed, thereby attracting more tourists". Presumably those reserves with more tourists attained a higher political profile. This political strength allowed the sites to argue for more budget allocation from government. Some sites also earned income from user fees. This study is important because it shows a strong and positive relationship between protected areas' budgets and tourism levels. Generally, those parks with high levels of tourism clients attain higher levels of political power. This power is then translated into higher budget allocations. It is important to recognise that substantial management budgets are necessary in areas of high usage to avoid excessive damage to the natural environment of the parks.

Parks often supply the most important part of the nature tourism experience, but typically capture little of the economic value of the stream of economic benefits. The low entry and use fees in parks are the result of many factors, one being the effort of a centralised budget allocation process in many governments. With this form of government financial management, the park management does not keep earned fees within its internal financial structure, and therefore sees little benefit in comprehensive fee collection. This budget process also contributes to a low emphasis on park visitor management. Such issues as return rates, length of stay, visitor satisfaction, and service quality all suffer when the financial return from the visitors is not tied directly to the financial operation of a park. This lack of emphasis on visitor management results in a dwarfed park tourism industry, and one where the visitors are often seen as being a problem, rather than a valuable asset. Under such a structure each recreational visit is a threat to management structure on a limited budget that cannot respond quickly to increases or changes in park use. Many governments see naturebased tourism as an important tool for economic development. Unfortunately, many have not invested sufficiently in staff training, infrastructure or park resources, or administrative structures that are needed to support nature tourism. This exposes sensitive sites to tourism-caused degradation.

Park Tourism: Planning and Management Issues

All national parks and protected areas have some level of visitor use. This can vary from just a few hundred to millions of visitors per year. Much of the visitor management is reactive rather than pro-active. The parks receive whatever visitor use that occurs, and then try to develop mechanisms to define and manage appropriate activities and levels of use. Often visitor management only takes place when some level of a problem is perceived. The parks may provide "take it or leave it" levels of tourism service. In other words, a type of recreation program or facility and a level of service is provided, with the visitor free to accept the service or to not participate. Traditionally visitors are not provided with ongoing visitor input mechanisms, such as public surveys. Visitors are expected to make their opinions about activities and services known only during management plan reviews or through complaints. However, increasingly park agencies are consistently and professionally monitoring and evaluating the needs, wants, and levels of satisfaction of their visitors. Parks Canada has even gone to so far as to establish a service quality goal for all units in the system. Each park or reserve is expected to provide services of sufficient quality such that visitors indicate high levels of satisfaction with those services using a standardised visitor monitoring methodology and measurement instrument.

Many park agencies are weak in tourism competencies. Those that are developed are usually the result of resource managers learning on the job about visitors and tourism management. Increasingly parks are gaining professional expertise in leisure pricing policy, tourism economics, marketing, tourism management, social statistics, service quality, and in leisure studies.

Parks Canada is one of the leaders in the development of high levels of competency in tourism management throughout the agency. This increase is stimulated by the need of the agency to gain operational income from tourism and the political realisation of the importance of a satisfied and mobilised constituency. In another example, Finland developed a unique visitor management approach with five components:

— different protected areas with different roles in relation to their recreational and educational services;
— a customer service chain model;
— standardised customer counts, surveys, and monitoring;
— a customer value creation process; and
— a customer service concept.

The lowest level of tourism competency typically occurs in park agencies where the central emphasis is on resource protection and the budget comes entirely from a central government pot. Whenever a park agency moves to a tourism-based budget where income from visitor services provides the income, there is a much higher emphasis given to tourism management.

Often the private-sector operators in and near parks have higher levels of tourism market expertise than do the parks themselves. In many parts of the world the private sector is the force behind tourism in parks. It is the private sector that attracts the visitors, services their basic needs, and provides all of the tourism services. A pointed example of this comes from Costa Rica, where the government agency responsible for the national parks and wildlife refuges has low tourism competencies. It is the private sector that is largely responsible for the internationally recognised, park-based ecotourism industry that has developed over the last 20 years. Significantly, the private, non-profit sector plays a major role in both reserve management and in tourism management in Costa Rica.

References

Boo, E. (1992) The ecotourism boom: planning for development and management. WHN Technical Paper Series Paper 2, World Wildlife Fund.

Driml, S., and M. Common. (1995). Economic and financial benefits of tourism in major protected areas. *Australian Journal of Environmental Management* 2:2, 19-39

Hornback, K.E., and P.F.J. Eagles. (1999). *Guidelines for Public Use Measurement and Reporting at Parks and Protected Areas.* Gland, Switzerland, and Cambridge, U.K.: IUCN.

IUCN. (1994). *United Nations List of National Parks and Protected Areas.* Gland, Switzerland, and Cambridge, U.K.: IUCN.

Goodwin, H. and Swingland, I.R. (1996) Ecotourism, biodiversity, and local development. *Biodiv. Conserv.* 5, 275–276.

Honey, M. (2002) *Ecotourism and Certification: Setting Standards in Practice*, Island Press.

8

International Ecotourism Initiatives

Ecotourism has been widely promoted as a viable alternative to ecologically and culturally degrading mass-tourism, as a form of sustainable development, which can yield immediate economic returns without risking major damages to local communities and the natural environment. However, a precise and unambiguous definition of ecotourism and subsequently a common understanding on its meaning and goals do not exist.

The problems of defining and conceptualizing ecotourism became utterly clear at an international conference entitled "Ecotourism: Concept, Design and Strategy", held in Bangkok from 6 to 8 February 1995. While there was general agreement that ecotourism is "nature-based, sustainably managed, inclusive of social and cultural aspects, and educational to tourists", only two of the speakers acknowledged that the concept includes "benefits to local people" and "involvement of local people" as components. Most ecotourism proponents were only concerned with tourism activities attracting visitors to natural areas and using the revenues to fuel economic development and, to some extent, fund conservation efforts; others used the term in a much broader sense and also stressed the "greening of mass tourism", i.e. by the reshaping of existing tours and hotel activities into more sustainable ventures.

Ecotourism and Question of "Sustainability"

Ecotourism policies, inspired by the 1987 Brundtland Report "Our Common Future" and later by the 1992 United Nations Conference on Environment and Development (UNCED), and their attempts at redefining development through "sustainability" and "justice" have been opposed due to the introduction of a dubious and false linkage between development and conservation. As such, they assist in concealing the actual dichotomy between ecological development and unsustainable economic growth.

Harrison points out that sustainability - and thus sustainable tourism - is an especially problematic concept when applied to socio-cultural phenomena:

> "Systems theory in social science has long experienced difficulty in distinguishing between social and cultural attributes that are functional and dysfunctional for the continuance of a social system and such analytical difficulties are unlikely to be eased by considering social systems as but a part of the wider natural environment, even though there may be evident a priori reasons for postulating such links. As a consequence, there are no clear criteria for determining which of the socio-cultural changes laid at the door of tourism represent an outright threat to a society's distinctiveness, which will help sustain the status quo, and which allow for adaptation in the face of external pressures to change. In fact, current differences in the assessment of the merits of social change in tourist-receiving societies are based more on differences in ideology and moral positions than on sound social analysis, the tools for which simply do not exist".

A major assumption underlying the ecotourism concept is that cooperation between the industry, governments, NGOs and local communities will translate into successful planning and implementation of ecotourism. This view can be rebuked as simplistic and unrealistic as it fails to take into account the greatly divergent interests and motives of the various participants in tourism. An ideal is being promoted where sustainable tourism can be achieved with all concerned institutions and their ethical values intact, while in reality, tourism development continues to rest with the narrow economic interests and action plans of the industry and other powerful institutions, which more often than not run contrary to cultural and ecological sensitivities of local communities. West/ Brechin recognise the strong tendency for ecotourism benefits to be monopolised by large-scale companies:

> "This pattern of centralised and highly capitalised tourism development is rooted in easy availability of capital, competitive advantage, power in the political economy, and compatibility with the interests of central governments in efficient 'shearing of the tourist sheep' to maximise foreign exchange.... Yet, [ecotourism] advocates continue blindly to promise

> economic cornucopia to local residents from tourism revenues without appreciation of the fact or understanding of the social structural conditions under which tourism can provide benefits to local people".

Under such conditions, it is easy to argue that discussions on conflict management and trade-offs between environmental and developmental goals, as suggested by some analysts, will hardly, if ever, bring about more equity and justice in tourism.

Without challenging the development process itself, "pristine environment", plant and animal species are primarily seen as commodities for tourist consumption, while significant social and political issues such as the maldistribution of resources, and inequalities in political representation and power are marginalised or ignored.

What has been often overlooked is the fact that ecotourism is a highly consumer-centred activity, mostly catering to the "alternative lifestyles" of the new middle classes of urbanised societies. Leslie, suggests that the norms, values and attitudes of Western industrialised societies need to change first:

> "Until we start tackling the issues, one of which undoubtedly is the impact of tourism, arising as a result of consumer societies, then sustainable tourism will never exist outside one's own backyard; but then that is not tourism!".

Ecotourism as Development

Claims among ecotourism protagonists persist that subsistence economies are non-productive, unprofitable and backward, and local people are ignorant of and responsible for environmental destruction. An effective ecotourism industry, it is argued, will stimulate economic growth and at the same time generate funds for conservation. There is hardly recognition of the fact that indigenous non-dominant, ecologically sound economic systems prevail in many places. Worse, ecotourism policies often involve the expulsion of people who in fact have contributed to shape biodiversity-rich landscapes and complex plant and wildlife habitats over generations.

A particular cause of anxiety is the fact that in relation to Third World countries, crucial questions regarding the international political economy and the widening gap between North and South are being sidestepped. Debt-ridden and increasingly tied to the global economy, many Third World governments in need of foreign exchange see little option but to exploit their natural resources for tourist consumption. Meanwhile, the global drive for biodiversity protection has enabled transnational corporations (TNCs) and organisations dominated by industrialised countries to increasingly intervene in Third World affairs and reap profits from their huge investments in ecotourism ventures.

Conservation-cum-tourism programmes such as the World Conservation Strategy, the Tropical Forestry Action Plan, Integrated Conservation and Development Projects (ICDPs), Global Environment Facility (GEF)-funded projects or the USAID-funded Biodiversity Support Programme have come under severe attack for their top-down approaches shaped by global techno-bureaucrats and their incapability to adequately coordinate activities with governments, NGOs and affected communities.

Globalisation and Liberalisation

The efforts of bi-lateral and multi-lateral development agencies such as USAID, UNDP, UNEP and financial institutions like the World Bank, all of which are significantly involved in influencing ecotourism agendas, are directed towards making Third World countries compliant through homogenisation of policies and standards. Examples are the Structural Adjustment Programmes and the liberalisation of trade and services through the General Agreement on Trade and Tariffs (GATT), which increasingly undermine economic and social progress in Third World countries and result in more destruction of cultural and ecological diversity.

"The General Agreement in Trade in Services or GATS, a separate agreement in the Uruguay Round of GATT, provides for the opening up of signatory countries to 100 per cent foreign investment in tourism services and disallows any protectionist measures for local tourism concerns. This will edge out small, independent enterprises as TNCs and their affiliates, with the advantage of financial resources and technology, muscle their way in to control the tourist trade in countries of the South. The GATT/ WTO [World Trade Organisation] and trends for liberalisation of tourism benefit not only the TNCs of the North. Joining the TNC bandwagon are a number of tourism-oriented development concerns in Taiwan, Hong Kong, Malaysia and Singapore which are independent or have linkages with Japanese capital and have expanded operations in the fast-growing East Asian region.

With the envisaged liberalisation measures under GATS rule (World Tourism Organisation 1994), tourism-related TNCs may in future face less restrictions to operate businesses across the world, without adequate mechanisms for public scrutiny and environmental auditing. As the ecotourism approach has neglected a wider debate on such crucial issues, there is a strong possibility that social and ecological conflicts around related programmes will aggravate

Environmental trap

Biodiversity and environmentally intact lands form the basis of ecological stability which has already been severely affected by industrialisation, urbanisation, unsustainable agricultural practices, and last not least mass-tourism. While ecotourism sounds comparatively benign, one of its most serious impacts is the usurpation of "virgin" territories - national parks, wildlife sanctuaries and other wilderness areas - which are then packaged as green products for ecotourists. With the tremendous expansion of commercialised ecotourism, environmental destruction - including deforestation, land deterioration, disruption of ecological life systems and various forms of pollution - has in fact increased and still carries on apace.

The contrast between conservation as practiced by local people, and profit-making conservation-cum-tourism ventures, is illustrated in the following example from Kenya: The Maasai from Loita Hills, some 320 kms south-west of Nairobi, have been fighting a fierce battle to prevent an indigenous forest from being turned into another ecotourism destination. As one local comments on the neighbouring Maasai Mara Reserve: "Tourism has been allowed to develop with virtually no controls. Too many lodges have been built, too much firewood is being used and no limits are placed on tourist vehicles. They regularly drive off-track and harrass the wildlife. Their tracks criss-cross the entire Maasai Mara. Inevitably, the bush is becoming eroded and degraded." Having experienced the negative effects of tourism by themselves, the community decided to protect the "Forest of the Lost Child", which for generations has been under their management and control and carefully kept as a sacred place for worship and communion with Maasai deities. The Maasai here are small-scale subsistence agriculturalists who have used resources in a sustainable manner. As a result, the forest is still dense and biodiversity-rich with abundant wildlife, plants which serve as herbal medicine, sufficient water sources and pastures to raise healthy cattle. Undoubtedly, not only the environment would be under siege, the spiritual and material way of life would also degenerate rapidly if tourism intrudes the area.

Searching for "untouched" or "authentic" places, young and adventurous travellers already served to open up many new destinations "off the beaten track" to mass tourism, accelerating the pace of social and environmental decay in host communities. Mainstream conservation groups have reinforced this trend by stressing the economic value of so-called conservation-cum-tourism projects, in order to make their environmental protection strategies attractive to international funding sources and the private industry in particular. As a result, even mega-resorts including luxury hotels, condominiums, shopping centers and golf courses, are increasingly established in nature reserves in the name of ecotourism. Such projects, in many cases protested as "eco-terrorism" rather than "ecotourism", tend to irretrievably wipe out plant and wildlife species and entire ecosystems to replace them by completely artificial landscapes.

The list of outright destructive developments is indeed endless. In Costa Rica, for instance, ecotourism has been a boom industry and a model for other parts of the world. If ecotourism can not succeed here, it is often argued, its role elsewhere looks distinctly doubtful. Meanwhile, however, local NGOs and small-scale tour operators have alerted to more and more unsustainable developments which clearly pave the way to mass tourism and turn the country into an "ecotourism graveyard". In 1994, Costa Rica's biggest tourism development ever, the Papagallo Gulf project on the North Pacific coast, was frozen due to charges of corruption and violation of legal procedures. The US$2 billion project, covering 2,000 hectares of pristine beaches surrounded by tropical forests, has set a target for 40,000 hotel rooms - almost five times as many as have been available in the entire country. The mega-resort has not only become a sensitive issue because of the alleged illegalities and devastating environmental impacts, but it has also unleashed a heated debate on Costa Rica's "sustainable" tourism future.

The countries of Indochina have only recently decided to develop their tourism industries, but there are already alarming signs that the very same mistakes as in other countries are being repeated. A case in point is Champasak province in Southern Laos, including the famous Lee Pee Waterfalls at Khon Phapheng on the Mekong River, which has been earmarked for international-level resort development. Some years ago, plans surfaced for the establishment of a US$300 million tourism complex, including luxury hotels, casinos, golf courses, a hydro-power station, and an international airport, on 500 hectares of land around the Lee Pee Falls, notwithstanding that the area has been proposed as a world heritage site because of its outstanding beauty and biological diversity. The Thai developer proposed to attract 1.8 million tourists annually to the area and claimed that the project would be carried out in an environmentally friendly manner in consultation with NGOs. However, the environmental report commissioned by the developer was strongly criticised as insufficient by environmentalists in Laos and the Thai press because the project involved deforestation, ecological disruption of the fragile Mekong river system, displacement of villagers, and probably undesirable social and cultural changes in nearby communities.

Indeed, commercial tourism to "unspoilt" pristine natural places - with or without an "eco"-prefix - is a contradiction in terms. To generate substantial revenue - whether for foreign exchange, tourism businesses, local communities or conservation -, the number of tourists has to be large, and that inevitably implies greater pressures on eco-systems.

Goodall agrees that the unprecedented growth of the tourism industry continues to threaten the environment, even where future operations are environmentally more efficient than current ones. He concludes that environmentally sustainable tourism will remain an elusive goal, not least because of its use of transport, air transportation for example:

> "British Airways's flying operations in 1992-93, when measured by standardised environmental performance indicators such as emissions per available seat kilometer improved over 1991-92, but total carbon dioxide and nitrogen oxides emissions increased by six per cent because of increased volume of business".

The issue of appropriate environmental management in tourist destinations has been addressed for many years now, but little has been done on the ground, and efforts to rearrange existing tourism towards ecotourism have rarely been successful. This is due to the general reluctance to limit tourism growth, the lack of controlled development and the lack of a thorough examination of the impact on the environment, i.e. the impact of resource utilisation, the consumptive nature of tourism and its continuous discharge of pollutants.

A number of environmental methodologies and techniques, such as ecological accounting, "polluter pays" principles, and environmental impact assessments (EIA) have been put forward in relation to ecotourism management, with little discussion and appraisal of their limitations and shortcomings. EIAs, for example, are usually not carried out by independent institutions, but private consultancies are hired for this purpose by the project owners who naturally want their schemes approved smoothly, without having to pay the high costs for precautionary environmental protection measures. Thus, EIAs are mostly produced in favour of the tourism companies rather than for the public good.

Benidorm in Spain, Acapulco in Mexico, and Pattaya in Thailand, all of which lost their reputation as fine seaside-resorts, clearly demonstrate the failure to clean up popular holiday destinations. Hawai'i is another typical example: The dark side of rampant tourism growth was showing up already 20 years ago, when descriptions of Waikiki began to surface as a "concrete jungle, a raucous sideshow - a billion-dollar cement mistake". Today, Hawai'i is encountering an even worse plague of problems which are by no means easily solvable. What islanders are witnessing is further over-building of their lands and continued environmental degradation characterised by loss of green space, beaches and marine life, overflowing sewers, depletion of limited water supply, and contamination of water and food by toxic chemicals. Hawai'i's tourism-centred economy, now almost completely dominated by powerful multinational corporations, has burdened locals with low wages, excessively expensive housing, and a cost-of-living 35 per cent higher than on the US mainland. The gaps between extremes of rich and poor has widened tremendously. Yet, political and business leaders believe that all these problems can be managed in an eco-friendly and democratic manner - even in prospect of a new tourism boom. Inspite of the alarming truths and far from effectively tackling existing social and environmental decay, there have been preparations for nearly doubling the number of tourists to the Hawai'ian islands, to 11.5 million annually, in the next 20 years. As a result of in-migrating workforce necessary to serve the growing tourism industry, the number of people residing in Hawai'i is expected to increase by 70 per cent!

"Self-regulation", as promoted by the "Green Globe" programme of the World Travel and Tourism Council's Environment Research Centre (WTTERC), has been lauded by industry leaders as a significant step towards environmentally sustainable tourism. However, in Josephides's words, "what generally happens is that a state monopoly, more often than not an inefficient one is replaced by a more ruthless, efficient private sector monopoly with short term aims, designed to satisfy city investors."

As a result of the worldwide trend to seek corporatisation or privatisation of basic services such as good roads, clean water supply, sewage treatment and waste disposal, local people will be left even more to their own devices, particularly in situations where poor sections of society are already disadvantaged. For instance, "private water companies will demand effective metering and cost recovery. In the Western Samoan capital of Apia, where tourism is targeted to play a major role in future development, there is the real possibility of a two-level water supply in the near future. The one being clean and expensive and available to a limited number of consumers (including tourist establishments) who can afford to pay for it, and the other, dirty and unreliable, and available to the majority."

Social Erosion

With the establishment of national parks, wildlife sanctuaries and other protected areas for conservation and tourism purposes, locals have in many cases lost their homes and livelihood, often without any compensation. McIvor explains the problematic relationship between people and nature reserves in Zimbabwe as follows:

> "(Colonial enterprise) appropriated the best agricultural lands in Zimbabwe for a small elite of European farmers and evicted communities, like some of the ones around Hwange, from their traditional homes to make way for recreational reserves and safari hunting areas that only benefited the small European population in the country or foreign visitors. This story is repeated in many of the park areas of Zimbabwe as communities contemplate the lost resources and opportunities on the other side of the fences that exclude them. While tourism may have brought some jobs and even stimulated a craft industry around a few of the parks, it is that sense of alienation and dispossession that has maintained uppermost in the minds of the people... For them, the best thing would be for the parks to disappear altogether, their animals destroyed and the foreign visitors transported to some other locations so that they could be left in peace."

Ecotourism planners and managers have in recent years put provisions for the "involvement of local communities" or "people's participation" high on the agenda of their project proposals, mainly as a means of confusing dissent and damage control. Yet, local residents have rarely been involved in the planning and implementation of ecotourism. Decision-making authority - including deciding whether a project should go ahead or not - has been generally denied. The independent evaluation of the GEF pilot phase, for example, states in relation to GEF-funded biodiversity projects, which often include an ecotourism component: "Most GEF work to date has been characterised by a top-down approach rather than responding to the needs of governments. It has not involved local communities in an effective way; it has sparked destructive competition among Implementing Agencies and other global organisations in the field of biodiversity; it has given inadequate consideration to sustainable use of biological resources; it has not meaningfully involved the NGOs; and it has been overly dependent on international consulting firms."

Koch observed in 1994 that in fact, "there is a growing awareness in South Africa that efforts to promote community participation in development projects can give rise to new forms of conflict and fragmentation", and recalls, for example, events in Maputaland, where the establishment of "community-based" game reserves has promoted secessionist organisations.

It does not come as a surprise then that ecotourism ventures are increasingly regarded as just another repressive form of economic development. Local communities continue to face exploitation and abuse, including the loss of cultural and social identity. It is a form of development that further undermines the autonomy of residents peoples by making them dependent on external forces and by making it harder for local organisations to function. As such, it erodes society's capacity and potential to sustain "self-reliance".

Other social issues such as changes of behaviour and values, especially among local youths, prostitution and AIDS, spreading fast from mass-tourism centers to ecotourism destinations, have been studiously avoided in ecotourism discussions.

Education and employment for locals have been emphasised as the main criteria for "involvement of communities" to make ecotourism projects successful. But the number of locals who can participate in tourism projects is relatively small, and in many cases, diverse social and economic activities are replaced by an ecotourism "monoculture", thus causing tremendous losses and leaving little choices for communities. Neither do local people necessarily benefit economically. Tourism-related employment has been greatly overrated, and locals are usually left with low-paying service jobs like tour guides, porters, food and souvenirs vendors, for example. In addition, tourism workers are not assured of year-round-employment and are laid off during off-season.

A study by Wells and Brandon on ICDPs confirms this bleak picture and asserts that "(t)he results (of ecotourism projects) thus far have been disappointing, to say the least. In general, all spending by visitors - on transportation, food, lodging, or even park entry fees - goes directly to the central treasury or to private corporate interests that have been granted concessions... At popular sites, tourism revenues greatly exceed protected area operating budgets. It is unusual for any of these revenues to be returned directly for park management and extremely rare for a revenue share to go to local people. For example, the value of visits to Khao Yai National Park in Thailand has been estimated at $5 million annually, which is about 100 times the national park budget; none of it goes to local people."

Even if locals participate in ecotourism in a meaningful manner, i.e. by setting up small-scale businesses, their livelihood is not secure as such initiatives can be easily jeopardised when financially strong outsiders take over ownership and control of the destination. In addition, large-scale ecotourism investments tend to drive up land and property prices as well as prices for daily necessities, so that higher incomes will be eaten up by inflation.

In Third World countries in particular, the economic benefits from ecotourism have been seriously questioned as it is abundantly clear that, as with conventional tourism, most of the profits are made by foreign airlines, tourist operators, and developers who repatriate them to their own economically more advanced countries. With increasing privatisation and deregulation of the global economy, there are now great and justifiable concerns that Southern countries will lose out even more. More liberalisation will lead to more foreign-owned tourist facilities and tour operations, and as a result, less income from tourism will remain in the local economy.

Indigenous Cultures Threatened

For ecotourism to claim that it preserves and enhances local cultures is highly disingenious. Ethnic groups are increasingly seen as a major asset, an "exotic" backdrop to natural scenery and wildlife. What has been in general ignored is the fact that the very same people have often been the targets of a consistent policy of suppression and exploitation by the dominant social groups in nation states. The simultaneous romanticisation and devastation of indigenous cultures is certainly one of the deepest ironies manifest in ecotourism.

There is rarely acknowledgement, and much less support, of indigenous peoples' struggle for cultural survival: the struggle for self-determination, freedom of cultural expression, rights to ancestral lands and control over land use and resource management. Participants of an Asia-Pacific Consultation on "Tourism, Indigenous Peoples and Land Rights", held at Sagada in the Mountain Province of the Philippines in early 1995, asserted the tourist industry affecting indigenous peoples as a reiteration of neo-colonial domination in the Third World. Their statement especially highlights the importance of land rights and land use practices:

"Indeed, it is through our worldview, in tune with our cosmic visions, our oneness with the land and our deep spiritual reverence towards the environment that our Mother Earth has few forests and rivers left. But these have become chief targets of the tourist industry under various euphemisms like 'ecotourism', 'sustainable tourism' and 'alternative tourism'. ... Unjust land laws must be repealed immediately. Ancestral lands are exploited in the name of public interest with no guarantee for resettlement. Colonial laws of the British rule are still in existence in India; the National Forest Act of 1927 outlaws Adivasis from their forests. American colonial laws in the Philippines are in existence through Presidential Decrees declaring Igorots (mountain peoples) squatters in their own lands. All over Hawai'i, the Kanaka Maolis are made strangers in their own lands by American laws. Elsewhere, Aboriginal Australians fight for native titles or possessions of lands arising from ancient tradition and customs and not from British laws. Similar situations exist among indigenous peoples in Papua New Guinea, Taiwan, Bangladesh and Nepal".

Whilst ecotourism attempts to fully integrate indigenous communities into the market-driven economic system, it keeps them as "archaeological" pieces to stimulate the tourists' nostalgic desire for the "untouched", "primitive" and "savage". Worse, irresponsible ecotourism promotion features photographs and descriptions of ethnic women, giving credence to the false notions that they are willing and available to be discovered by tourists. Apart from resisting to take-overs of ancestral lands by tourism developers, indigenous peoples organisations and support groups have strongly denounced ecotourism which has produced "human zoos", as such practices abuse human dignity and involve socio-economic and cultural disruptions which amount to ethnocide.

Hence, ecotourism often appears to replicate, if not intensify, dysfunctional cultural impacts which it has claimed to be countering, and tends to reinforce racial and class-bound institutionalised discriminatory processes which indigenous peoples have been exposed to since colonial times. As Munt puts it:

> "Racism is not only institutionalised and commodified, but it is reproduced and fashioned through the media, so that we increasingly consume false images of what we are supposed to value, whether of the native hill tribes of Northern Thailand or Mayan Indians in Guatemala. We are no longer able to transcend or look beyond the 'smiling faces', or challenge this image-reality... Widescale repression of human rights, deeply rooted racism and intense class struggle are null and void in the brave new world of adventure tourism."

Legitimising Discourse of Ecotourism

Ecotourism proponents have been keen to present a positive and optimistic picture for the environmentally concerned public. However, the rhetoric used in literature and conferences in support of ecotourism creates considerable confusion between ecotourism as a theoretical ideal, what has been planned and what has been actually achieved. The terms "opportunity" of and "potential" for ecotourism are found in abundance to give the impression that this activity has already proven to function well.

Some proponents have also portrayed ecotourism as a "potentially" powerful tool to boost economic development and conservation and at the same time conceded that it has the "potential" to destroy natural and cultural resources. In an attempt to shield themselves from well-founded criticism, they assert that ecotourism is not a cure but a good start, because, under "well-managed" conditions, it "can" produce positive results. Whatever conflicts occur, the message is: The benefits of ecotourism development will eventually outweigh the problems, and there is no other alternative than to promote it.

Another example for manipulating perceptions through language is the focus on "local participation" as it gives the idea that ecotourism programmes are capable of empowering marginalised communities and effecting positive changes in local development. This is in fact another neutralising technique, as it directs concerns away from an analysis of economic and political structures which are counter-productive to local political organisation and development options that put people's and ecological concerns at the forefront.

Tourist centres in the Asia-Pacific region, which have turned out as environmental nightmares and are, thus, increasingly shunned by tourists, are now refurbished with the well-sounding term "mature resorts". A paper recently circulated by PATA suggests: "A key to maintaining or enhancing a mature resort's viability is to attract new niche markets through redevelopment.". In addition, it has become common practise that public relations specialists assist governments and the tourism industry to dilute "negative" press reports on tourist destinations by presenting "positive" counter-information - concerning environmental problems as well as political crises, human rights violations, crime, tourist rip-offs, prostitution and AIDS - and, thus, to change public opinion and attitudes that may affect tourism growth.

With the industry's global environmental programmes and strategies in place and being aggressively promoted, tourism market leaders have begun to hide themselves behind the facade of self-imposed regulations such as "ethical" codes of conduct and environmental management systems, although "they are quite aware that self-regulation will not address the basic problems."

Whereas there has been substantial research and analysis on the critical dimensions of ecotourism, the concerned industry, government agencies and mainstream conservation groups have hardly considered or encouraged such efforts. This may be due to the fear that a focus on adverse social and impacts and risks might create or reinforce image problems, and, subsequently, lead to cuts of funding and a slow down of the commercialisation of the lucrative ecotourism industry.

Inspite of the adoption of popular environmental slogans and consideration of local capacities, ecotourism projects usually offer "single-factor solutions", as if it was to develop tourism or nothing. Legitimation and credibility is mainly sought from professionals - state officials, consultants, investors, businesspeople, tour operators, etc. - who often follow the same narrow and short-sighted approach and are prone to vested interests. Meanwhile, grassroots-oriented initiatives, which have developed from outside of official and professional circles and are based on a broader and more reflective analysis of social and environmental realities, are rejected as unscientific and unverified.

Considering the already sufficient evidence of the serious adverse effects, the huge investment of financial, human and infrastructural resources for ecotourism promotion appears to be misplaced and irresponsible. There are signs that now even greater amounts of money are being spent in expectation to make up for previous mistakes and losses, thereby increasing the risks further. In addition, it is hardly justifiable that overwhelming amounts of resources for such highly commercial and risky ventures like ecotourism are channeled away from other more holistic research and projects that can contribute to more sustainable and realistic solutions to pressing social and environmental problems.

Policy Proposals

Driven by Western environmentalism, a new Green Revolution in the form of ecotourism, is sweeping the Third World, along with market leaders' efforts directed towards the removal of all barriers to travel, including: physical barriers, economic barriers, organisational barriers and legal barriers. The above facts and deliberations lead to the conclusion that ecotourism creates more losers than winners and the impacts on society and nature will take disastrous proportions, especially in poor countries.

The ecotourism lobby, predominantly based in Northern countries, has in recent years exercised tremendous financial and political influences over official, academic and NGO-circles in developing countries to achieve their goals.

For many years already, alternative forms of tourism have been discussed, aimed to at least reduce the negative impacts on nature and society in destinations. Some empowered communities have experimented with small-scale, locally controlled and sustainable tourist activities by themselves, while rejecting development impositions on their lives; a few projects of this kind have actually operated with some successes. Yet, all these initiatives have certainly not posed a real challenge to the status quo.

If it is agreed, that development strategies should be genuinely people-oriented, "bottom-up" and aimed at a redistribution of resources and wealth, then the first priority is to acknowledge that tourism, including its "green" offshoots is, compared to probably all other activities, fraught with images and myths. This means there is a clear need to expose the harsh realities, clarify the contradictions, and confront decision-makers with the multi-faceted problems involved. Critical analysis and assessment, which may vary from case to case, are indispensable in this discourse. The study of local resistance related to ecotourism policies and projects and radical movements formed by national and international advocacy and campaigning groups offer revealing perspectives on these issues. Public education and awareness building also remain keys to productive discussion and creative re-thinking concerning tourism options. How else can local communities make a reasonable and responsible decision as to whether to make their homes and lands a playground for ecotourists or not?

Particularly in this age of globalisation, there is indeed a strong case for governments in developing countries to restrain such activities imposed from outside, at the very least until there are sufficient capacities to effectively scrutinise, monitor and control developments through administrative and legal mechanisms, and informed public debate. The establishment of these frameworks will also involve the need for a profound redistribution in decision-making and for devolution of proprietorship and management of resources to local communities.

Likewise, an adequate infrastructure for participatory research and public education should be set up on all levels that takes on the task to review and re-evaluate ecotourism and develop accountability mechanisms to phase out unsustainable policies and projects. Simultaneously, expanded and adequate resources should be made available for holistic and integrative studies of other fields of development aimed to bring about alternative solutions to tourism and the diverse problems resulting from urbanisation, industrialisation and over-exploitative agricultural land uses.

Governments and international agencies should facilitate an open and public review of ecotourism concepts and audits of existing ecotourism projects. Illegalities committed in the name of ecotourism - including the encroachment of public lands and protected forests, violation of local and indigenous customary rights, diversion of natural resources, and corruption - must be investigated and prosecuted by responsible government agencies.

Finally, tourism has to be addressed as an extension of wasteful and unsustainable consumerist lifestyles of affluent societies. In this context, more efforts need to be made to fully inform and educate tourists on the adverse environmental and social impacts of ecotourism. Regulations and laws should be enacted to prohibit the advertising of unsustainable ecotourism projects as well as promotional materials which project false images of ecotourism destinations and are demeaning to local and indigenous cultures.

International Year of Ecotourism

The United Nations General Assembly, by its resolution A/RES/53/200, proclaimed the year 2002 as the International Year of Ecotourism. The 7th session of the Commission on Sustainable Development (CSD-7) held in New York in April 1999 gave a mandate to the World Tourism Organisation (WTO) and the United Nations Environment Programme (UNEP) to assume responsibility for the IYE, and adopted a resolution inviting Governments, the UN and NGOs to organise activities in preparation for the IYE (International Year of Ecotourism). The text proclaiming 2002 as the International Year of Ecotourism referred specifically to Agenda 21 adopted in Rio in 1992.

All states were invited to contribute to the IYE by discussion papers and own activities, in order to arrive at a comprehensive understanding of ecotourism as a contribution to sustainable development, and in particular for the less developed countries. In the preparation of and during the International Year, WTO and UNEP aimed at involving all the actors in the field of ecotourism, with the following four objectives in mind:

i) Generate greater awareness among public authorities, the private sector, the civil society and consumers regarding ecotourism's capacity to contribute to the conservation of the natural and cultural heritage in natural and rural areas, and the improvement of standards of living in those areas.
ii) Disseminate methods and techniques for the planning, management, regulation and monitoring of ecotourism to guarantee its long-term sustainability.
iii) Promote exchanges of successful experiences in the field of ecotourism.
iv) Increase opportunities for the efficient marketing and promotion of ecotourism destinations and products on international markets.

The Official Launching Ceremony of the International Year of Ecotourism 2002 took place at the United Nations Headquarters in New York on 28 January 2002. The event was attended by the UN Deputy Secretary-General, Louise Fréchette, WTO Secretary-General, Francesco Frangialli, UNEP Executive Director, Klaus Töpfer, Simone de Comarmond, the Seychelles Minister of Tourism and Transport who chaired the session, Leticia Navarro, the Mexican Minister of Tourism, and Megan Epler Wood, then President of The International Ecotourism Society. It also involved the participation of several government ministers, representatives of intergovernmental organisations, members of the diplomatic corps accredited to the UN, and representatives of leading industry associations and non-governmental groups.

Initiative by World Tourism Organisation

The World Tourism Organisation undertook various activities in preparation for and during the IYE at different levels. These activities are the following:

WTO Recommendations

In line with the first three objectives, WTO recommended its 139 Member States, in September 2000, to undertake activities at the national and local levels, such as:

a) define, strengthen and disseminate as appropriate, a National Strategy and specific programmes for the sustainable development and management of ecotourism;
b) provide technical, financial and promotional support for, and facilitate the creation and operation of small and medium size firms;
c) set up compulsory and/or voluntary regulations regarding ecotourism activities, particularly in what refers to the environmental and socio-cultural sustainability;
d) establish national and/or local committees for the celebration of IYE, involving all the stakeholders relevant to this activity;
e) inform the WTO Secretariat of the activities planned for 2002 requesting, if appropriate, whatever support they deem necessary.

During 2001 and 2002, more than 50 Member States sent information on their national activities, published in the WTO IYE 2002 website.

Regional Conferences

Ten regional WTO conferences were successfully organised between March 2001 and April 2002 to exchange experiences, examine problems, promote cooperation nationally, regionally and internationally, and identify future challenges. These conferences were structured according to the same four themes of the World Ecotourism Summit:

— *Theme 1*: Ecotourism planning and product development: the sustainability challenge
— *Theme 2*: Monitoring and regulation of ecotourism: evaluating progress towards sustainability
— *Theme 3*: Marketing and promotion of ecotourism: reaching sustainable consumers
— *Theme 4:* Costs and benefits of ecotourism: a sustainable distribution among all stakeholders.

These preparatory meetings were held in Mozambique for Africa, in Brazil for the Americas, in Kazakhstan for CIS countries, in Austria for Europe, in Greece for Mediterranean Europe, the Middle East and North Africa, in the Seychelles for island destinations, in Algeria for countries with desert areas, in the Maldives for the Asia-Pacific region, in Moscow for Russia and neighbouring countries and in Fiji for South Pacific islands. In total, over 3,000 stakeholders, representing public sector tourism and environmental authorities, non-governmental organisations, ecotourism businesses, academic institutions and independent experts, participated in these preparatory meetings, where some 200 case studies were presented.

WTO also organised, jointly with UNEP, a Web-Conference on Sustainable Development of Ecotourism during April 2002, in which nearly 1,000 people from 88 countries participated. The prime objective of the conference was to provide easy access for a wide range of stakeholders involved in ecotourism to exchange experiences and voice comments, especially for those who had not been able to attend the regional preparatory conferences that had taken place previously.

In order to contribute to the preparation of the World Ecotourism Summit, UNEP organised or participated in preparatory events involving over 3,000 stakeholders. These meetings were held in India for NGO and grassroots organisations (with Ecological Tourism in Europe), in Belize for Central America (with The International Ecotourism Society—TIES), in Seychelles for Small Island Developing States and other small islands (with WTO), in India for South Asia (with TIES), in Peru for South America (with TIES), in Thailand for Southeast Asia, (with TIES), in Kenya for East Africa (with TIES), in Sweden for the Arctic Circle (with TIES), and in Austria for Europe (with WTO). The experiences and results from all these regional meetings and the Web-Conference were used as a base of discussions at the World Ecotourism Summit in Quebec, Canada on the 19-22 May 2002.

The World Ecotourism Summit

The Summit was the principal event to mark 2002 as the International Year of Ecotourism, and the culmination of one-and-a-half-year long preparatory process. It was an initiative of the World Tourism Organisation (WTO) and the United Nations Environment Programme (UNEP). It was hosted by Tourism Québec and the Canadian Tourism Commission. It was successfully held in Quebec City, Canada from 19 to 22 May, 2002, with the participation of 1,169 delegates from 132 different countries, representing public, private, NGO, academic and research institutions, intergovernmental, national and international development and aid agencies, as well as local and indigenous communities and individual experts. Among the participants, there were 30 Ministers of State and senior officials from WTO, UNEP, UNDP, UNCTAD, UNESCO, CBD, ILO, GEF, Inter-American Development Bank and the European Union.

The main outcome of the Summit is the Quebec Declaration on Ecotourism, a document that was prepared through wide consultation at the Summit and contains general guidelines, as well as stakeholder-specific recommendations for the sustainable development of ecotourism. The final version of the Quebec Declaration and a comprehensive Final Report, including the summaries of the regional, thematic and stakeholder-specific sessions of the Summit, was published in print and is also available in the internet. These publications have been submitted to the World Summit on Sustainable Development (WSSD) held in Johannesburg, as official documents. More than 10,000 copies of the Quebec Declaration have so far been distributed at numerous events, including WSSD.

Other Activities

In line with objective 4, the World Tourism Organisation took part in various fairs and published a set of market studies on ecotourism.

Ecotourism activities at Tourism and Trade Fairs: Besides all the preparatory conferences and seminars and the World Ecotourism Summit, WTO participated in special ecotourism activities or workshops at various tourism and trade fairs:

— *Reisepavillon (Hanover, January 2002)*: WTO, jointly with the German Technical Cooperation Agency (GTZ) convened the Forum International with the participation of public authorities, ecotourism companies and experts. WTO and GTZ subsidised the participation of more than 50 small suppliers of ecotourism products and services from 20 developing countries.

— *FITUR (Madrid, February 2002)*: WTO organised a special Session on Ecotourism for Latin American Member States. FITUR, since its beginning, represents a meeting place for public and private organisations of Latin American countries. Ecotourism development and management is a key factor to assure a sustainable development of tourism in Latin America, and this region is a major supplier of ecotourism products and destinations for the European outbound ecotourism market. The special relevance of this meeting was also confirmed by the high number of tourism professionals (approximately 200) who participated in it.

— *International Adventure Travel and Outdoor Sports Exhibition (Chicago, February 2002)*: The IATOS exhibition is a specialised tourism event that attracts professionals and public interested in the adventure travel, ecotourism, outdoor sports and other related tourism activities. WTO supported the event, participated as exhibitor and delivered a presentation at a parallel ecotourism conference.

— *ITB International Tourism Fair (Berlin, March 2002)*: In connection with the International Year of Ecotourism, 2002, the 2002 edition of ITB included a special "Sustainable Travel Exchange—Travel with Sense" exhibition, at which WTO participated. It featured a wide range of initiatives and tourism attractions that emphasise environmental and social sustainability. Visitors found initiatives by international tour operators and smaller suppliers too, as well as model tourism products from over 50 countries.

— *EXPO-ECOTURISMO 2002, Ecotourism Exhibition and Trade Show (10-11 September 2002, Caracas, Venezuela)*: WTO supported the event, participated as exhibitor and delivered a presentation on the results of the World Ecotourism Summit.

— *Ecotourism Market Study Series*: In order to evaluate the trends and size that characterise the main ecotourism generating markets, the World Tourism Organisation published seven pioneer country reports on the following markets: Canada, France, Germany, Italy, Spain, UK and USA. This is a comparative study; therefore, in each country a common definition of ecotourism and a similar methodology have been used.

Other Special Publications for the IYE

(a) *WTO publications* are:

— *Compilation of Good Practices in the Sustainable Development of Ecotourism*: 55 case studies from 39 countries, presented in a systematic form, drawing lessons that can be applied at other destinations. It is the second volume within the series of Good Practices published in the area of Sustainable Development of Tourism. The sustainability aspects are further detailed according to specific elements of ecotourism such as: conservation, community involvement, interpretation and education, as well as environmental management practices.

— *Guidelines for the Sustainable Development and Management of Tourism in National Parks and Protected Areas (revised edition, jointly with UNEP and IUCN)*: These guidelines aim to build an understanding of protected area tourism, and its management. They provide both a theoretical structure and practical guidelines for managers.

— *International Ecolodge Guidelines (WTO sponsored and contributed to this publication by The International Ecotourism Society)*: This book incorporates the latest research and techniques available, providing guidance from the day that an idea is conceptualised to the time when the ecolodge is built and operating. This book is an invaluable resource for anyone associated with ecolodges or other nature-based accommodations and facilities

— *Final Report of the World Ecotourism Summit*: This publication contains the summaries of the preparatory conferences, the conclusions of all Summit sessions, the Quebec Declaration on Ecotourism and other technical documents presented at the event.

— *Compilation of Good Practices in Small Ecotourism Businesses*: Published in 2003, this third volume within the WTO series of Good Practices contains 64 case studies from 47 countries.

(b) UNEP publications are:

- The UNEP Manual for the International Year of Ecotourism, containing orientation for interested parties to collaborate with the Year. The Manual has been posted on UNEP website.
- A double issue of the "Industry and Environment review" on Ecotourism, including articles presenting successful ecotourism experience from all parts of the world.
- A handbook: "Ecotourism: Principles, Practices and Policies for Sustainability", with basic background data and references for governments and practitioners, jointly produced with TIES.
- A CD-ROM, published jointly with WTO, distributed at the World Ecotourism Summit, with all preparatory conference reports, keynote addresses and 72 presentations from delegates.

WTO Website

A special page has been created on the WTO Web-site for IYE related activities: http://www.world-tourism.org/sustainable/IYE-Main-Menu.htm

This page, regularly updated, contains all the material regarding the activities related to the IYE.

WTO Events

WTO has also supported several international, regional and national ecotourism events during 2001 and 2002, in which WTO representatives delivered technical presentations. In the national conferences that have taken place after the Summit (in Sri Lanka, Bulgaria, Hungary, Spain, Venezuela, Rumania, Brazil, Portugal) WTO has assisted governments in adapting the Quebec Declaration on Ecotourism to the particular conditions of each country, and thus helped to implement the Summit's recommendations.

Other Global Projects Linked to the IYE

- The Convention on Biological Diversity (CBD), as part of its mandate within the international programme on sustainable tourism development under the Commission on Sustainable Development process, has developed a set of international guidelines on biodiversity and tourism development. The draft guidelines, together with a booklet containing their main features, were presented at the World Ecotourism Summit in May 2002. Afterwards, the guidelines have been tested through three case studies for their applicability and effectiveness. The draft guidelines will be transmitted for adoption, as well as recommendations for future work to be undertaken on this issue, to the seventh meeting of the Conference of the Parties, which will take place in Kuala Lumpur in April 2004.

— The World Travel and Tourism Council (WTTC) has been engaged in various initiatives that are linked to ecotourism; such as the Corporate Social Leadership in Travel and Tourism, and the Tourism Industry Report prepared for the World Summit on Sustainable Development, among others.

— The Ecotourism Databank, established by the Centre for Sustainable Tourism at the University of Colorado, supported by UNEP, WTO and by the US Forest Service, is a searchable online library on ecotourism, containing relevant documents and presentations on ecotourism.

— The Sustainable Tourism Stewardship Council, managed by the Rainforest Alliance through the support of the Ford Foundation, is proposing a global accreditation body for sustainable tourism and ecotourism certifiers.

— The World Legacy Awards managed by Conservation International and the National Geographic Society.

— The World Wide Fund for Nature published in July 2001 the Guidelines for community-based Ecotourism Development.

National Policy

Before and during the International Year of Ecotourism, numerous countries organised special events, put in place new policies and developed inter-sectorial and international co-operation, and many other activities. The information for the following analysis has been obtained through a specific worldwide survey conducted by WTO among National Tourism Authorities between January-May 2003, as well as from the communications of WTO Member States reporting on ecotourism-related activities throughout the IYE and the WTO Tourism Market Trends 2002 survey, which included a specific section on ecotourism aspects.

IYE national committees

Following the recommendations of WTO, 47 national and local tourism authorities established committees for the celebration of IYE, involving all the stakeholders relevant to this activity; 14 in Africa, 12 in the Americas, 4 in the Middle-East, 5 in Europe, 6 in East Asia/Pacific and 6 in South Asia. Most of these committees include similar stakeholders at different levels, such as:

— National Tourism Authorities, normally as initiator and president of the Committee
— Other ministries or government agencies: principally for culture and heritage, environment, foreign affairs/co-operation, economy/trade/development or the forest department, but also for communication, agriculture, decentralisation, scientific research, biodiversity, meteorology, mines and geology, and sport.
— Foundations or NGOs (environment, ecotourism, development, co-operation)
— Private Sector: tour operators, travel agents, hotel owners and their trade associations
— Local or regional authorities (mainly tourism officials)
— Tourism boards
— National bodies for protected areas
— Local community representatives
— Environmental or ecotourism associations
— Universities / academicians
— Tourism trade organisations
— Hotel trade national federations or chambers
— National representatives of international conservation organisations (e.g. Nature Conservancy, Conservation International, WWF)
— Media

a) Some committees include also other stakeholders like national social economic councils, intergovernmental organisations, funding institutions, tourism observatories, national funds or councils for tourism promotion, national indigenous agencies, etc.

b) Some countries without national ecotourism committees for the IYE, developed other mechanisms to undertake ecotourism activities during 2002.

In general, a large majority of the IYE national committees were established to develop and organise activities for 2002. However, in order to support ecotourism development in the future, most countries decided to maintain their committee and pursue its activities. Some countries, like Azerbaijan or the Czech Republic, have not set up any national ecotourism committee so far, but are planning to establish one in the future.

National strategy for ecotourism

While the establishment of national ecotourism committees aimed at undertaking activities in this field, setting up national strategies for ecotourism reflects a deeper involvement of these countries toward ecotourism. 48 countries defined, or are currently defining, a national strategy/plan for ecotourism development in their territories (8 in Africa, 12 in the Americas, 8 in East Asia and Pacific, 13 in Europe, 3 in the Middle East and 4 in South Asia). There are of course differences in the way these strategies are established in these countries:

— Around a third of them have clearly developed a separate strategy for ecotourism, focusing on different aspects.

— Some others have included ecotourism as one of the main segments of a more general Tourism Strategy (Rwanda, Haiti, Cambodia, Republic of Moldova, Kazakhstan, Ecuador, Sweden or Uruguay, for example), of the tourism chapter within the National Economy Development Strategy (Lithuania), of the Nature Tourism Strategy (Portugal) or of a Rural Tourism Strategy (Morocco).

Ecotourism policies in some countries (Spain, Georgia, Argentina) are mainly managed by bodies in charge of protected areas. Some countries (Syria) make reference to an expansion of green areas or to a greening programme for tourism areas (Mauritius). Others (Bangladesh, Egypt, Jamaica, Oman) have no national plan but refer to regional or local strategies or programmes for ecotourism. Other countries (Madagascar, Lebanon and Panama) are not currently setting up any ecotourism strategy, but will present a project of such strategy in the future. Yet, other countries (Bolivia, Russia, and Hong Kong SAR) have a general strategy for sustainable development of tourism, which applies to ecotourism issues as well.

Activities and events

In order to contribute to the international debate on ecotourism during the

IYE, many national governments organised events at the international, national, regional or local levels. The regional distribution of these events worldwide is quite homogenous, even if Africa, Americas and Europe have witnessed more activities. Typical activities undertaken include:

— National or regional congress, seminar or workshop
— International seminar or conference
— Launching of ecotourism project or programme
— World Tourism Day celebration on ecotourism
— Launching ceremony for the IYE
— Exhibitions (photo, painting, etc.)
— Sport events (e.g. cycling or trekking tours)
— Local workshop or celebration
— National or international fairs
— Conference in national parks and other protected areas (including inauguration of a new national park)
— Regional, national or international festivals (food, music, traditional arts, etc.) related to ecotourism
— Ecotourism or nature tourism award ceremony (national or regional)
— National public debate
— Good practices contest

Participants represented all types of stakeholders, including other international or intergovernmental organisations like WTO, TIES, IUCN, UNEP, UNESCO, etc. The themes frequently associated with ecotourism in these activities were: mountain tourism, rural tourism, agrotourism, protected areas, community-based tourism, cultural tourism, traditional handicraft conservation, nature conservation, ecolodge, sustainable tourism, poverty alleviation, etc. The IYE was also the occasion to develop partnerships: Some events were celebrated jointly with the International Year of Mountains 2002, and a few of them were co-organised by two bordering countries (i.e. Guinea with Senegal).

Besides, many States, whether they organised ecotourism activities or not, have mentioned that they participated at conferences and seminars prepared by other countries, and also at the World Ecotourism Summit in Québec. They also supported many events organised by NGOs and associations. Some countries mentioned they already plan to organise other ecotourism events in 2003 or beyond.

Publications

States produced various types of publications in the framework of the International Year. According to the information sent, Europe has been the region with more written outputs, followed by the Americas. As it could be expected and in view of the numerous activities undertaken world-wide, final reports and other outputs (announcement, declarations, charters, etc.) represent an important part of the publications produced during the IYE. Other IYE outputs published by different countries around the world can be summarised as follows:

- Inventory of the principal ecotourism sites, facilities and attractions
- Publication of national ecotourism guides (included sometimes in a more general tourism guide)
- Publication of national or regional ecotourism strategies, plans or programmes (as a whole or summarised)
- Guidelines on ecotourism development (focussing on the specificity of the country: desert areas, wetlands, etc.) for communities, for local bodies or for business operators i.e. ecolodge
- Guidelines for tourists on responsible tourism (i.e. on diving, trekking, adventure, etc.)
- Publication of the proceedings of national ecotourism conferences, seminars and workshops
- Special promotional posters, CD-ROM, videos (on national parks for example), Web sites, etc.
- Articles on ecotourism in ecotourism or related magazines or in the national press
- Market research on national or regional ecotourism markets (also on rural tourism)
- Regional maps with ecotourism attractions (among others)
- Publication on ecotourism development in national parks
- National speeches presented at the WES or other international conferences
- Ecotourism impact studies

The IYE was the occasion for some governments to commission special studies on ecotourism (market, feasibility or consultancy studies at national level or for certain regions, etc.) to be published in the near future. They also supported universities' publications, like thesis dealing with specific aspects of ecotourism development. Some of the IYE outputs from governments were prepared in order to be distributed at the World Summit on Sustainable Development in Johannesburg.

Tourism authorities (ministries or tourism boards) were usually associated with other bodies for these publications, sometimes through the national committees they have formed: universities, research centres, other ministries (economy, environment, culture), national parks authorities, nature associations, institutes of geography, etc. Ecotourism was often dealt in association with other themes like environmental protection, agrotourism or rural tourism, mountains, poverty reduction, etc., thus demonstrating its many linkages.

Stakeholders Participation and Support

The United Nations system in general, and the World Tourism Organisation in particular, has always encouraged national authorities to develop participative mechanisms in their activities. The creation of national committees for the IYE was certainly a most important tool for this purpose: many countries mentioned it as the main consultative organ on ecotourism questions. Nevertheless, more actions regarding stakeholders' participation were conducted before and during the International Year. Some countries (Botswana, for example) made extensive efforts to ensure that all stakeholders, government agencies, communities, NGOs and the private sector were given the opportunity to express their opinions at each phase of the preparation of the national strategy/plan for ecotourism.

Participation of communities is considered not only necessary in ecotourism policy construction, but as its main justification. Workshops, intended for regional or local authorities and indigenous/local communities, seem to be quite regularly used, while a few countries, like Georgia and Iran, used a mail consultation. In some cases, workshops are only organised at local levels for the dialogue between local authorities and their communities. Otherwise, these local bodies and communities are invited to participate in national or international events to express their opinion (events can be specially designed for enforcing upward flow of information from civil society to national authorities, like in Mexico for example with the "Ecotourism Community-based Enterprises National Encounter" or in Australia, where the Aboriginal Tourism Australia was one of the main IYE partners). Nevertheless, consultation about ecotourism questions can be more restricted according to the context and can be limited to:

— Contacts with experts and NGOs at national level
— Consultation of local communities for a specific local ecotourism project. As a result of this consultation are that local communities are highly involved in the project management. In some countries, the execution of an ecotourism project needs the approval of the local communities.
— Consultation with indigenous and other stakeholders in national reserves
— Trans-ministerial consultation
— Contacts with private sector
— Dialogue with indigenous national associations
— Support to NGOs and private foundations that work with communities
— Part of a consultancy or feasibility study on ecotourism development

Some countries (like Puerto Rico) recognised that such consultative mechanisms are still limited, but are in a process of strong improvement. Others, like Thailand, say that consultative bodies have been formed, but have not properly been used so far. In general, very few States consider that regular democratic mechanisms are enough for ecotourism questions.

One of the recommendations made by WTO during the IYE was for the National authorities to provide technical, financial and promotional support for, and facilitate the creation and operation of small and medium size firms. Indeed, ecotourism has to be built rather on a dense network of small businesses than on the big concentrations that often characterise the global tourism market. States, principally those from the Americas and Africa, which seem to consider tourism in general, and ecotourism in particular, as a strong component of their economies, responded to this recommendation using different methods:

Technical assistance to small enterprises and local communities (advice, feasibility studies, capacity building, etc.) comes before a direct financial support, which often is directed to local bodies and businesses working for specific ecotourism projects or destinations (i.e. focused on protected areas). Since some national governments, like in Botswana, did not have enough funds to finance projects, they facilitated contacts between ecotourism project developers and development partners (NGOs, funding agencies, banks or donors). They also developed administrative and market facilities for local ecotourism businesses, for example, setting up positive discrimination for national businesses, fiscal exemption, loans with low interest and other commercial incentives (provided businesses comply with some conditions on environment or community participation).

A few countries have set up national funding programmes for tourism enterprises (like in Malaysia, or for young entrepreneurs in Greece). Others have organised national financing plans for small businesses in general, within a sustainable development programme (for equipment and environmental improvement, training and research, etc.). These global plans also benefit ecotourism small size firms. For example, in Greece, in the framework of the Operational Programme for Competitiveness, the Ministry of Development provided financial support for "Infrastructures for the attraction, management and targeted information dissemination for visitors" (pathways, observation posts, etc).

Besides, 2002 was also the occasion to start new programmes focused on nature tourism or ecotourism: in Indonesia, for example, the Ministry responsible for tourism has already planned for 2003 to give financial and technical support for communities who lived around ecotourism destinations. States like Saudi Arabia, Seychelles, Bangladesh and Yemen mentioned they have not undertaken support actions so far, but they plan to develop such mechanisms. As far as promotion is concerned, ecotourism businesses could enjoy certain advantages from national promotional campaigns, especially when these focused on the natural heritage, like the one in Chile for example.

In the framework of these campaigns, national tourism authorities or boards directly issued promotional material (postcards, posters, logo, etc.). An interesting experience was undertaken in Hungary, where the government organised study tours in national parks for the media. Besides, part of the technical and financial support accorded to small businesses described previously, was dedicated to promotional purposes (Web-sites creation, publication of advertising brochures, setting up of ecotourism business networks, ecotourism stand in foreign fairs, shouldering costs for the participation at IYE events, etc.).

Awareness

Around 20per cent of reporting countries mentioned the setting up of global awareness raising campaigns on ecotourism or responsible tourism. These campaigns took place sometimes within a national ecotourism promotion operation or in a general campaign on environment. Besides, in these or other countries, awareness actions for local operators were conducted, principally focused on environmental training for tourism employees and guides, by means of special workshops or leaflet distribution (including the Québec Declaration on Ecotourism).

Similar processes have been used for local communities, especially those living within or close to protected areas (training, guidelines for ecotourism business development, etc.). Schools and universities were also subject to particular attention from some governments, which set up special information campaign for teachers and students. Special attention has also been paid to state employees (police officers, national parks' agents, etc.) and to elected members through awareness meetings. For the general population, the following actions were undertaken: environmental public march, cycling or hiking (Hong Kong SAR), clean up campaigns (mountains, beaches,), tree-planting activities, special IYE lottery draw (Ecuador), free guiding services on ecotourism sites (Sri Lanka), open days in national parks for the World Tourism Day (Hungary), etc.

Governments also supported or facilitated the work undertaken by associations (for environmental protection, responsible tourism or community-based tourism). In general, various means were used to spread the information: leaflets, posters, maps, videos, radio or TV programmes, exhibitions (photography contests in Chile, exhibition on environment friendly technologies in Jamaica), ecotourism festivals, special pages in Web-sites, documentation during flights (Colombia), national or foreign specialised press (special supplements on protected areas), edition of special stamps (Senegal), code of conduct for tourism and environment (Cyprus, in preparation in Dubai), etc.

Regulation

Legislation

Various countries set up national strategies in order to plan ecotourism activities. Nevertheless, a scarce minority of them voted special legislation on ecotourism. Only three countries, from the ones that reported their IYE activities to WTO, mentioned law texts concerning ecotourism: Ecuador, Puerto Rico and Philippines (executive order on ecotourism). A few others (Colombia, Mexico, Cambodia, Spain and Bangladesh) have legislation mentioning the concept of ecotourism. However, around 20per cent of reporting countries declared they foresee the preparation of such legislation.

Furthermore, there is one case of trans-border legislative document for ecotourism activity in the Caucasus region currently being negotiated between Azerbaijan and Georgia. Besides, many countries without specific ecotourism legislation stated that ecotourism activities are regulated by various existing acts and laws. Firstly, national laws on environmental protection or national parks ordinances determined tourism and ecotourism development and management. Secondly, some countries already have tourism legislation that includes natural heritage protection and land planning, even a specific law on sustainable development of tourism (e.g. Czech Republic and Republic of Korea). This legislation can deal with forms of tourism close to ecotourism, like agrotourism and rural tourism (Cyprus, Uruguay, Morocco), nature tourism (Portugal) or adventure tourism (Chile).

In general, governments associate ecotourism to legislations concerning the following themes: archaeology, land use planning, national Agenda 21 (Jordan, Tunisia), hunting, architecture, forest, water, natural reserves, agriculture, wetlands, cultural heritage, wildlife and coasts.

Certification systems and indicators

A voluntary approach to regulate ecotourism activities is through developing or encouraging the use of indicators, certification schemes and/or ecolabels. Very few countries have set up specific national ecolabels for ecotourism businesses. Four have reported such fact to WTO: Kenya, Ecuador, Sweden and Thailand. Regarding ecological or quality labels and awards for general tourism, the list is a little longer: Morocco (on rural tourism), Colombia, Costa Rica, Estonia, Peru, Austria, Malta, Spain and Maldives. Other countries mentioned that some regional labels exist on their territory and that some efforts are being made to encourage the adoption of international ecolabels already existing (e.g. Blue Flag, Green Globe). Existing certification schemes concern mainly catering facilities and accommodation.

They focus principally on environmental issues (energy and water saving techniques, noise and air quality and the use of eco-friendly products and materials), but also on social ones (local community awareness) and services quality. It is to be noticed that around half of the countries that do not have any certification schemes so far intend to develop such mechanisms in the future. This reveals a strong awareness among tourism national authorities regarding the high potential of voluntary schemes for the regulation of ecotourism activities. Other countries, like Lebanon and Samoa, have developed a national system for sustainable tourism indicators that is applied for monitoring ecotourism activities as well.

Cooperation

Many countries collaborated in the preparation of or participated in international seminars on ecotourism before and during 2002, notably with WTO. Besides, some governments have enforced the link with international organisations or with other national governments in ecotourism programmes. One out of three countries that informed on their IYE activities stated they collaborated with international organisations or programmes for ecotourism development purposes.

Besides the World Tourism Organisation, other organisations mentioned are the following: UNESCO (MAB Programme), UNDP, UNEP, the United Nations Development Fund for Women (UNIFEM), The International Ecotourism Society (TIES), The World Conservation Union (IUCN), WWF, United Nation's Economic and Social Commission for Asia and the Pacific (ESCAP), the Pacific Asia Travel Association (PATA), European programmes (INTERREG), Convention on Biological Diversity, International Bird Watching Association, Conservation Tropical RARE, Adventure Travel Society, Caribbean Tourism Organisation (CTO), The New Partnership For Africa's Development (NEPAD), Emigration International, the South Asian Association for Regional Cooperation (SAARC), BIMST-EC (Bangladesh, India, Sri Lanka, Thailand Economic Cooperation) and International Mountaineering Organisation.

Similarly, one out of three countries have established an international collaboration agreement related to ecotourism with other national governments (including between national parks authorities) for research, training programmes, ecotourism regulation, etc. This collaboration is mainly in the case of border countries, which often share protected areas, like Lesotho and South Africa, or Ecuador, Peru and Colombia. Nevertheless, it can also be the result of a technical co-operation between non border states (like between Costa Rica and Chile or Ecuador for example, for the adaptation to ecotourism of the tourism certification scheme).

Co-operation can be included within a cross-boundary general development plan: for example, the Amazonian Strategic Plan or Euro-regions Development Plans, like the project for the Carpathian Mountains, which involves 7 countries. This form of co-operation arises from the need to alleviate the negative consequences, for tourism development among others, of states and nations being separated by frontiers, and to counteract the tendency for these areas distant from their capital cities to be neglected. Many countries also benefit from the technical or financial assistance of national co-operation and development agencies: AECI (Spain), GTZ (Germany), Belgian, Irish, USAID, Canada Green Fund, New Zealand or international funding agencies or organisations: Inter-American Bank, World Bank, European Union, African Development Bank, Asian Development Bank, GEF, etc.

WTO's Follow up Activities

i) Based on the results of the survey, a detailed publication on exemplary government practices will be prepared, containing documents and further nalysis of the main IYE results in ecotourism policy, planning, stakeholder participation and support, marketing, regulation, etc.

ii) The collection and dissemination of good practices is being continued after the IYE: A compilation of good practices in small ecotourism businesses has been published in April 2003

iii) Support for small and medium size ecotourism companies will be provided for the application of the IYE recommendations, through the above-mentioned good practice compilation, seminars and other dissemination tools, as well as through taking an active role in the preparations for the UN-designated International Year of Micro credit.

iv) A series of ecotourism seminars for protected area managers will be organised, starting with two sub-regional events in Africa.

v) WTO is continuing its work in the field of certification systems for sustainable tourism. It supports the process of creating the Sustainable Tourism Stewardship Council, a global accreditation programme. The Organisation recently prepared a set of recommendations for governments for supporting or creating national certification systems. Besides the general procedure and the criteria recommendations, that are fully applicable to ecotourism operations as well, specific criteria have been included for ecotourism and suggestions have been made to facilitate the participation of small tourism businesses in certification programmes. Based on these recommendations, a series of regional seminars will be organised in the course of 2003 and 2004.

vi) A series of national seminars on Local Agenda 21 programmes has been initiated in 2003. The Seminars involve principally municipal governments and tourism officials and promote the application of integrated tourism management systems.

vii) The work on sustainable tourism indicators is also continued, with a new international study expected to be published by early 2004. The resource book and manual will include specific sections on the application of indicators at natural sites and ecotourism attractions.

viii) A comprehensive Destination Management Network is currently being developed, which will specify the main types of tourism destinations, including natural and ecotourism destinations.

UNEP's Follow up Activities

As a follow-up to the International Year of Ecotourism, UNEP will:

i) Develop guidelines and recommendations for successful eco- and sustainable tourism policies and strategies. The guidelines will be based on the results of 6 regional multi-stakeholder workshops in 2003 and 2004. In 2004, UNEP will publish a handbook for governments based on this consultative process, including an action plan for next steps in setting global standards, and a proposed financial structure for implementation of the action plan.

ii) Develop tools to support the implementation of the guidelines, such as:

   a) a report on the links between eco- and sustainable tourism development, conservation of biodiversity and poverty alleviation called "Mapping Tourism Footprints" (with Conservation International and the International Institute for Environment and Development), with a series of GIS-referenced maps overlapping tourism investments, conservation hotpots and social variables such as the UNDP's Human Development Index.

b) guides for small- and medium tourism enterprises and local authorities in developing countries, on renewable energy, waste management and Local Agenda 21 processes.

iii) Continue implementing demonstration projects such as the UNESCO/ UNEP "Linking Tourism and Biodiversity Conservation in Six World Heritage sites". In 2003, the project will focus on training participants in public use of parks, ecotourism and nature interpretation, consolidating and disseminating lessons learned through publications, and fostering public-private partnerships. The Indonesian sites will begin activities this year.

Ecotourism Policies

Policies are dynamic in nature reflecting the evolving understanding of the society. Policies may need periodic revisions based on emerging perspectives and techniques. Hence policy formulation benefits from guidelines brought out at different times, at different governance levels and by different agencies. The policy being recommended here, also finds its origin in existing practices, policies and guidelines both within and outside the country. Nevertheless it is realised that ecotourism is a vast business domain having wide repercussions on the socio-ecological fabric of an economy. Therefore, it is time to develop the guidelines into clearer regulations, strategies and action plans. While guidelines can be national, policies need to reflect state-specific needs and imperatives of ET. Hence these recommendations provide space for regions and communities to innovate and to retain their ecological and cultural integrity. In terms of the process of policy formulation, there has to be dialogue, consensus, bottom-up approach, decentralised governance and redressal mechanisms. The Quebec declaration (2002) says:

> *"formulate national, regional and local ecotourism policies and development strategies that are consistent with the overall objectives of sustainable development, and to do so through a wide consultation process with those who are likely to become involved in, affect, or be affected by ecotourism activities;"*

Following this process, we propose national and state level institutions and mechanisms to promote and sustain ET in the country as a tool for conservation, livelihood and development.

Comprehensive guidelines for planning and monitoring community based ET with case studies can be found in WWF, 2001. ET related policies in five countries and one region are briefly reviewed here.

Bhutan

With its mighty mountain ranges and relatively undisturbed ethnicity, Bhutan is a natural destination for ET. Bhutan's new long term ET policy (previous ET policy was formulated in 2001) from Tourism Ministry continues its emphasis on 'high value low impact' tourism and tries to enhance community participation and conservation capacities. For a large functional democracy as India, high value low impact may not be an ideal slogan because of its trade-off with local economic and educational benefits.

Nepal

Ministry of Culture, Tourism and Civil Aviation (MoCTCA) is responsible for policy, planning, licensing, regulations and overall monitoring of the tourism industry. Nepal Tourism Board (NTB) conducts planning, research and product development. Country's National Ecotourism strategy and marketing programme of 2004 emphasises cross-sectorial cooperation at national planning level, more private participation, poverty alleviation through promotion of village tourism and a full-fledged marketing programme. Other government agencies in Nepal's ET sector include National Planning Commission and Department of National Parks and Wildlife Conservation. Nepal plans to set up a new body under NTB to coordinate ecotourism development, stressing on community participation in decisionmaking, planning and participatory techniques.

Insights for India from Nepal's ET sector include involvement of the departments of Culture and Civil Aviation and the need for a supporting research and training establishment.

Thailand

Thailand used well-marketed, open, mass-tourism for economic recovery from the Asian crisis of the last century. This resulted in large-scale interference with local ecology and Thai culture. Thus from 1995-96, Thailand started focusing on sustainability and in 1997 set up National Ecotourism Councils comprising of representatives of the public, academic, private, and NGO sectors. The purpose of these councils was to oversee the development of National Ecotourism Policy and Action Plan as well as to appoint subcommittees on various aspects of ecotourism management. National Ecotourism Action Plan is a five-year implementation plan for the period 2002 to 2006 with details of the projects needed to be implemented. Thrust areas include Tourism resources and environment management, Education and awareness building among the public, Cooperation among local people, Marketing promotion and tour guide, Basic infrastructure and ecotourism services development and Ecotourism investment support and promotion.

The broad-base structure of ET councils may be suited to Indian context. This could help the state institutions of varied nature in timely revisions and provide space for consultations on international agreements and other national policies impacting the sector. Among the new thrust areas of Thailand, public awareness to provide quality Ecotourists and stress on cooperation among and within communities are India's concerns too. An implication to be drawn from this constitutional monarchy is the dependence on international tourists leaving conspicuous cultural footprints.

Sri Lanka

Heavy dependence on highly fluctuating flow of international tourists and the absence of national ecotourism policy leave the ET sector in Sri Lanka unable to grow beyond the occasional ecotours and ecolodges organised by mass-tour operators. Spice/plantation walks (agroforestry/ botanical tours) are becoming popular ecotourism activity, helps save many indebted planters of this island economy.

Lessons for India from Sri Lankan experience include the scope of farm tourism and risks associated with putting too many eggs in an industry far too sensitive to insecurity and instability. Leakages in terms of proportion of receipts from tourism flowing out of the economy (for Sri Lanka, leakages are around 27% makes the over dependence on international tourists a not so viable proposition. Experience of Thailand and Sri Lanka implies that quality domestic tourists would be more valuable to ET because of low cultural impact, higher domestic educational value and more predictable demand.

Australia

'Ecotourism Australia' is an incorporated not-for -profit association of lead industries with diverse members. It develops certification Program along with Cooperative Research Centre for Sustainable Tourism. Australian certification is accepted as 'International Ecotourism Standards' in 2001. 'Green Globe 21' (a global Affiliation, Benchmarking and Certification program for sustainable travel and tourism) has exclusive license for the distribution and management of the International Ecotourism Standard. An important insight from Australian policy is the freedom for aborigines to choose the way they wanted to be involved: as employees, investors, or participants in preventing negative impacts.

Africa

Wildlife tourism generates almost one third of Kenya's foreign exchange earnings. In South Africa there has been spectacular increase in number of visitors to game and nature reserves. The integrated policy on cultural or village tourism of the African region was formulated in the line of policy on Community National Parks Management. This provides an example for the South Asian Region, with comparable context of culture and history. African policy recognises that 'Low-volume' need not be 'low-impact' if quality tourism is not ensured by regulations and awareness. African policy sets the precedence of participatory identification of criteria and indicators. Involving stakeholder communities, they have adopted a broad set of criteria and a complementary set of model-specific indicators from WTO.

Many developing nations including India have made commitments under the WTO's General Agreement on Trade in Services (GATS). The scope of GATS extends to all member governments, which includes national and local governments.

India

Table 1 provides an overview of the strengths and weaknesses of existing policy documents related to ET in the country at national and state levels. The policy defines and hence approaches ET with a clear conservation bias. It lays out cardinal principles suggesting the importance of involvement of local communities, minimising the conflicts between livelihoods and tourism, environmental and sociocultural carrying capacities. It also perceives that ET should be part of integrated development of the area. It emphasises the role of careful planning in infrastructure development and a detailed benefit cost analysis prior to implementation. It talks about standards, continuous monitoring and codes of conduct for visitors.

Table 1. Status of existing policy initiatives in the ET sector, India

	India	Karnataka	Kerala		
	Ecotourism in India: Policy and Guidelines 1998	Wilderness Tourism Policy 2003	Kerala Tourism Eco initiative 2004	Conservation and Preservation of Areas Act 2005	Participatory Ecotourism programme of Forest Dept 2005
Objectives	Same as general tourism; to be a unifying force, preserving natural and cultural heritage	Opening up forest areas for ecotourism	Eco-Certification Scheme to make each sub-sector within the tourism sector to be eco-friendly	Avoid unsustainable tourism	Promotes participatory Ecotourism programme through EDC/VSS
Strengths	Identified key players in Ecotourism	A beginning	One of the first standards issued by a state, intends to cover all sectors of tourism	Regulates unplanned growth of tourism	Inter-departmental cooperation, financial assistance for establishing community ecotourism
Weakness	No institutional set-up, fiscal incentives or community ownership.	Public sector gets priority/Monopolized access to forest areas	Less stress on socio-cult-ural aspects including local employment, insufficient incentives for achieving the standards	In conflict with the expected powers and rights of 'panchayathi raj' institutions	Difficult procedural formalities involving many sanctioning authorities

	Sikkim	Madhya Pradesh	Himachal Pradesh
	Sikkim wildlife (Regulation of Trekking) Rules, 2005	Eco & Adventure Tourism Policy 2002	Policy on Development of Ecotourism 2001
Objectives	To regulate trekking activities in the state	Opening up forest areas for eco/adventure	Opening up of forest areas for community based ecotourism
Strengths	Details penalties for offences and rewards for reporting offences	A beginning	Community involvement, well detailed institutional set up
Weakness	Confines only to trekking activities	Stresses on adventure tourism and private sector. Lack of community involvement	Too much stress on trekking and less on other ecotourism activities

The way to go ahead is to develop these guidelines into action plans, incorporating the missing components: institutional support, monitoring criteria, incentives and regulations.

Purpose of this comparative table (Table 1) is to bring out the need for broad comparability of policies across states, even allowing for contexts to determine this to a great extent (mountains, coast, wildlife rich, etc.). There is a clear picture of the confusion that surrounds the states' definition of ET and their articulation of what they want to achieve—ET as purely business proposition in MP, privileged access to public sector in Karnataka, to detailed institutional set up in HP and enabling decentralisation in Kerala and Sikkim. Identified policy gaps in Indian ET sector are the following:

Policy Gaps

— *ET policies often conflict with policies of the Tourism sector*: Tourism policies promote infrastructure development and recommend simplification of environmental regulations to attract large capital inflows. ET objective is to minimise new infrastructures and comply with all environmental regulations.

— *Role of Government and other institutions not specified*: Except in Kerala, there has been no attempt to craft a nodal agency for ET, where the roles of different government departments are specified in any policy document.

— *Conservation—heavy and local benefits not emphasised*: Due to the conceptual ambiguity in defining ET, most of the stakeholders, except forest department have not taken a pro-active role. The result is an apparent conservation bias at the cost of local stakes.

— *Lack of community involvement*: Even in cases where local community is projected as beneficiaries, benefits mostly confine to employment of a few locals as guides and cooks. Other forms of benefit sharing are nearly absent in the sector. Even the employment benefit could be higher if capacity building was considered as a prerequisite for assessing the eligibility of locals for different jobs.

— *Lacks clear and measurable indicators to monitor*: Except for self imposed regulations in specific activities (e.g.: picking up waste thrown around while trekking, checking the number of plastic articles etc.) there are no clear guidelines for monitoring even for environmental impact. Sociocultural parameters so far remain ignored.

— *Absence of links between Monitoring and Regulations or incentives*: Wherever some regulations exist, they are not linked to continuous monitoring. Well-laid out incentives linked to regular monitoring and standards can streamline the connectivity between practices and policy.

Way Ahead

Objectives of ecotourism cannot be met without a focused and concerted approach. The identified gaps are interconnected and consequence of the absence of a commonly accepted definition of ET. National policies and guidelines should be drawn for the proposed components of ET, within the purview of international environmental treaties and related Indian legislations, incorporating equity and fair Trade principles. State wise regulatory institutions and regulations can be based on these guidelines but should reflect grassroots ecological and cultural integrity. While all environmental legislations apply to these enterprises, the sector cannot sustain without targeted regulations, as it's potentially significant impacts on environment and social fabric.

Extensive decision making powers of Panchayati Raj Institutions under schedule XI of constitution could be made use of to ensure realisation of all aspects of ET: nature and culture conservation, generating livelihood opportunities and regular monitoring. These constitutional rights need to be taken into account for negotiations under international agreements.

Coordinations and Enforcement for Ecotourism

While NED can be largely a strategic and co-coordinating agency, SED needs to be pro active and enforcing. SED will be responsible for streamlining the process involving registration, monitoring, certification and incentive mechanisms. The dichotomy between mass-tourism and ET needs to be clarified by SED in distinctive terms taking the local natural and cultural thresholds.

Any one/society/ company/ foundation wanting to use the title ecotourism should be registered under respective model category with the SED. SED could be decentralised to help inclusive and speedy processing, in a phased manner till district level. Decisions (registration of a proposed ET venture, certification, incentives, monitoring etc.) should be taken by a representative committee with officials from concerned departments in the proposed destination (eg: pollution control board, water authority, revenue, land use, agriculture, horticulture) and concerned community representatives.

The directorate would facilitate the realisation of the concept of ET i.e. of nature and culture conservation, education and local ownership.

Certain areas with unique ecology or ethnicity could be declared by SED as 'ecotourism zones' to imply that all tourism activities in that locality will be confined to Ecotourism. Initiative to declare ecotourism zones can spearhead from the co-coordinating agency or any concerned agency/person and should be facilitated by the ET directorate in pursuing the processes involved.

Registration

Once SED receives a proposal for an ET venture, it needs to set up a decisionmaking committee for assessing the feasibility of the proposal, involving all concerned stakeholders. Therefore the first step envisaged in processing a proposal for ET registration is to identify the stakeholder groups. Later, in a time bound process, the proposal should be evaluated with necessary site visits, stakeholder consultations, and scientific assessment of potential ecological, sociocultural and economic criteria under the guidance of the Institute of ET. Cut off scores for certain indicators, based on location specific features and weightage is a participatory decision of the management committee in consultation with IET. Based on the multi criteria assessment of any proposed ET enterprise, if the ecological and sociocultural criteria score with respect to the cutoff limit, the proposal could be recommended by the decision making committee for registration to the ET director as an ecotourism enterprise. Registration fees could be differentially charged: nominal for community owned enterprises and highest for private corporate sector.

Monitoring

The approach to monitoring as in the case of assessment should be integrated and based on multiple criteria so as to cover the triple bottom line of ecotourism. Multicriteria assessment/ monitoring is needed for registration, certification or incentives. Enterprises can use the monitoring procedure for self-evaluation. Indicators for which attaining a high score is practically difficult due to absence of systems or technology in place (eg: solid waste management) should be taken up by the SED with concerned authorities for improving the system. SED will be responsible for monitoring the growth of the sector as a whole and to judge carrying capacity along with the Institute of ET at regular intervals. SED should discuss the guidelines issued from time to time by the international Ecotourism society and NED and the rating criteria prescribed by various agencies to formulate needed indicators and regulations in the local context.

Certification

Allows inclusion of an enterprise as per request in a specific category of rating (eg: five star or three or two stars) which should be linked to careful monitoring by the decisionmaking committee and IET, based on selected parameters and indicators. This is proposed to link ratings to charges and monitoring based incentives. Voluntary compliance based ratings already existing in the field may not fully achieve this. These ratings should be indicators of the prices charged. The categories can be identified with range of cut off scores attached to each model in each of the three sets of criteria elaborated (ecological, sociocultural and economic). Ratings should follow a process: application to the SED, setting up a monitoring committee for site visits and valuation of the criteria. The committee will meet and discuss the scores under each of the three criteria and decide on granting certification with respect to cumulative rating with inputs form IET.

Incentives

— Considering the societal benefit of ecotourism, these enterprises should be eligible for income tax relief. This incentive should be differential as minimum relief to the corporate and maximum to the community owned initiatives. The incentives could be linked to the scores under social benefits especially to the indicators of employment and multiplier benefits. Determine and utilise measures of direct and indirect societal benefits to provide differential fiscal incentives to models.

— Investment cost in complying with selected input intensive parameters (eg: rain water harvesting, non-conventional energy etc.) could be extended as green loans under a subsidised lending rate. SED could recommend such lending schemes to the banking sector, based on the scores under economic criterion.

— Operational eco-friendly inputs (like insecticides, fertilisers and water purifiers of organic origin or reviving endangered skills using local natural assets) should be considered fro-green subsidies to all registered ecotourism enterprises.

Capacity Building

The SED should facilitate capacity building through the IET for people involved in running the enterprises as also in sensitisation of public on responsible travel and visits.

References

Ministry of Tourism. (1998). *Ecotourism in India - Policy and guidelines.* Government of India

Patton, C. V. and D. S. Sawicki. (1993). *Basic Methods of Policy Analysis and Planning.* Prentice Hall.

The Ecotourism Society. (1993). *Ecotourism guidelines for nature tour operators.* N. Bennington, Vermont: The Ecotourism Society.

Vivanco, L. (2002). *Ecotourism, Paradise lost - A Thai case study*. The Ecologist. pp. 32(2):28–30.

WTO. (2002). Report on Sustainable Development of Ecotourism. Web-Conference, International Year of Ecotourism Organised by UNEP and WTO.

Ziffer, K. (1989). *Ecotourism: The uneasy alliance.* Washington D.C.: Conservation International.

Bibliography

Anderson, D.L. (1993). A window to the natural world: The design of ecotourism facilities. In *Ecotourism: A guide for planners and managers, Volume 1,* K. Lindberg and B. Hawkins (eds.), 116-133. N. Bennington, Vermont: The Ecotourism Society.

Batta R.N.(2000). *Tourism And The Environment: A Quest for Sustainability.* Indus Publishing Company, New Delhi. pp. 203.

Baumol, W.J., and W.E. Oates (1977). *Economics, environmental policy, and quality of life.* Englewood Cliffs, New Jersey, USA: Prentice Hall.

Boo, E. (1990). *Ecotourism: The potentials and pitfalls.* Volumes 1 and 2. Washington D.C.: World Wildlife Fund.

Brandon, K. (1996). *Ecotourism and conservation: A review of key issues.* World Bank Environmental Department Paper No. 033. Washington D.C.: The World Bank.

Cater, E. (1994). Cater, E., and G. Lowman. ed. *Ecotourism in the Third World — Problems and Prospects for Sustainability in: Ecotourism, a sustainable option?.* United Kingdom: John Wiley and Sons.

Ceballos-Lascurain, H. (1996). *Tourism, Ecotourism and Protected Areas.* Gland, Switzerland: IUCN.

Department of Tourism. (2005). *Conservation and Preservation of Areas Act.* Government of Kerala.

Ethos Consulting. (1990). *Adventure Travel in Eastern Canada.* Ottawa, Ont.: Tourism Canada.

Goodwin, H. and Swingland, I.R. (1996) Ecotourism, biodiversity, and local development. *Biodiv. Conserv.* 5, 275–276

Groombridge, B. & Jenkins, M.D. (2000). *Global Biodiversity: Earth's Living Resources in the 21st Century.* Cambridge, UK: World Conservation Monitoring Centre.

Hawkins, D., M. Epler Wood, and S. Bittman. (1995). *The ecolodge sourcebook for planners and developers.* N. Bennington, Vermont: The Ecotourism Society.

Hawkins, D.; Lamoureux, K.; & Poon, A. (2002). *The Relationship of Tourism Development to Biodiversity Conservation and the Sustainable Use of Energy and Water Resources.* Report to the United Nations Environment Programme.

Holden, A. (2000). *Environment and Tourism.* London: Routledge.

Honey, M. (2002) *Ecotourism and Certification: Setting Standards in Practice*, Island Press

Hornback, K.E., and P.F.J. Eagles. (1999). *Guidelines for Public Use Measurement and Reporting at Parks and Protected Areas.* Gland, Switzerland, and Cambridge, U.K.: IUCN.

IUCN. (1994). *United Nations List of National Parks and Protected Areas.* Gland, Switzerland, and Cambridge, U.K.: IUCN.

Kamauro, O. (1996). *Ecotourism: Suicide or Development? Voices from Africa #6: Sustainable Development, UN Non-Governmental Liaison Service.* United Nations News Service.

Kinnaird, M.F. and O'Brien, G. (1996) Ecotourism in the Tangkoko DuaSudara Nature Reserve: opening Pandora's box? Oryx 30, 65–73

Lindberg, K. and Enriquez, J. (1994) An Analysis of Ecotourism's Economic Contribution to Conservation in Belize. Vol. 2, Comprehensive Report, World Wildlife Fund and Ministry of Tourism and the Environment (Belize)

Mandip Singh Soin (ed). (2004). A ready Reckoner for the tourism Industry. *Ecotourism and Environment Handbook.* Ministry of Tourism, Govt. of India.

Mehta, H., A. Baez, and P. O'Laughlin. (2002). *International Ecolodge Guidelines.* N. Burlington, Vermont: The International Ecotourism Society.

Ministry of Tourism. (1998). *Ecotourism in India - Policy and guidelines.* Government of India

Orams, M. B. (2000), Tourist getting close to whales, is it what whale watching is all about? *Tourist Management,* 21(6): 562-569.

Patterson, C. (2002). *The business of ecotourism.* Rhinelander, WI.: Explorer's Guide Publishing, pp.40.41, Business Evaluation Criteria.

Potter, C; Cohen, J.; & Janczewski, D. (Ed.). (1993). *Perspectives on Biodiversity: Case Studies of Genetic Resource.* Washington, DC: American Association for the Advancement of Science.

Sanders, E. and E. Halpenny. (2001). *The business of ecolodges: A survey of ecolodge economics and finance.* Burlington, Vermont: The International Ecotourism Society.

The Ecotourism Society. (1993). *Ecotourism guidelines for nature tour operators.* N. Bennington, Vermont: The Ecotourism Society.

Vivanco, L. (2002). *Ecotourism, Paradise lost - A Thai case study.* The Ecologist. pp. 32(2):28–30.

Weaver, D. B. (1998). *Ecotourism in the less developed world.* C.A.B. Int. Pbl. 288 pp.

Ziffer, K. (1989). *Ecotourism: The uneasy alliance.* Washington D.C.: Conservation International.